Modern Mathematics in the Light of the Fields Medals

Michael Monastyrsky
Institute of Theoretical and
Experimental Physics
Moscow, Russia

A K Peters, Ltd.
Wellesley, Massachusetts

Dedicated to the memory of my mother

Editorial, Sales, and Customer Service Office

A K Peters, Ltd.
289 Linden Street
Wellesley, MA 02181

Library of Congress Cataloging-in-Publication Data

Monastyrskiĭ, Mikhail Il'ich.
Modern mathematics in the light of the Fields medals / Michael Monastyrsky.
p. cm.
Includes bibliographical references (p. —) and index.
ISBN 1-56881-083-0
1. Mathematics—History—20th century. 2. Fields Prizes.
I. Title
QA26.M66 1996
510—dc20 96-22137
CIP

Printed in the United States of America
01 00 99 98 10 9 8 7 6 5 4 3 2 1

Contents

Foreword

Freeman Dyson
Institute for Advanced Study
Princeton, New Jersey

This little book is extraordinary in at least three ways. First, it contains in only 150 pages a synopsis of the whole of modern mathematics. Second, it is fully international in scope, giving equal emphasis to outstanding achievements of mathematicians of all countries. Third, it comes with a superb bibliography, giving references to original papers and review articles from which the expert or non-expert reader can obtain detailed information about the many subjects that are only briefly mentioned in the text.

The idea of summarizing the whole of mathematics in 150 pages seems at first glance to be absurd. Probably the apparent absurdity of it is the reason why nobody has done it before. Monastyrsky was brave enough to attempt it in spite of the absurdity, and he has brilliantly succeeded. To succeed, he needed not only boldness, but also a rare ability to read and understand a vast literature in many languages. He needed breadth of vision to see connections between ideas in widely separated contexts. He needed good taste in order to simplify without distortion. Monastyrsky may not be unique among mathematicians in possessing the necessary knowledge, vision, and taste to succeed in this task. He is unique in combining these qualities with the will power to push the job through to completion.

What Monastyrsky has produced is a road map to the territory of mathematics, with the Fields Medals as a convenient set of nodal points. Like a road map, this book consists mostly of names and connections. It names the important people and the important concepts, and sketches briefly the important connections between them. It is not a topographic map, describing in detail the beauty of the landscape. It is impossible in 150 pages to explain all the names or to follow all the twists and turns of the connections. To be useful, a road map must not be overloaded with information. I believe this book will be useful, both to experts and nonexperts, in giving us a quick overview of mathematics. It tells us how to move into unfamiliar parts of the territory without getting lost. And it tells us, through the abundant references to the bibliography, where to look for more detailed information when we need it.

The human genome project gives us another metaphor for the usefulness of this book. The genome is, like mathematics, a complicated structure stretching the limits of human understanding. The genome project consists of two parts, sequencing and mapping. Sequencing means discovering the precise sequence of bases out of which the entire apparatus of human genes is built. Mapping means identifying the most important genes and finding out roughly where they sit in relation to one another. Sequencing is still a long-range goal, while mapping is already a practical tool. So far as practical applications in science and medicine are concerned, mapping is more useful than sequencing. Mapping gives us a road map of human genes, while sequencing will give us the genes themselves in minute detail. Applying this metaphor to mathematics, we may compare the sequencing of the genome to the monthly publication of *Mathematical Reviews*, now filling several hundred massive volumes, while the mapping of mathematics is done quite adequately by Monastyrsky's 150 pages.

Monastyrsky had the misfortune to live a large part of his life as an "internal exile" in the Soviet Union, unable to travel freely or to publish freely. That misfortune, however painful it was to Monastyrsky personally, had useful consequences for the writing of this book. One consequence was that he had more time for omnivorous reading than mathematicians in happier circumstances enjoy. Another consequence was that he was familiar with the work of many outstanding mathematicians whose names were hardly known outside the Soviet Union. This book is, among other things, a poignant memorial to the two generations of Soviet mathematicians whose lives and works were stunted by the vagaries of the Soviet regime.

A final word of instruction to the reader. For a road map to be useful, it is not necessary for the reader to understand the significance of the place-names. If you dip into this book at random, you may well feel like Alice on her journey through the looking-glass, when she was confronted with the poem "Jabberwocky" and spelled out the words:

'Twas brillig, and the slithy toves
Did gyre and gimble in the wabe;
All mimsy were the borogroves,
And the mome raths outgrabe.

Alice remarked that the only thing she could understand in the poem was that somebody killed something. So you may find that some of Monastyrsky's pages sound like "Jabberwocky," full of unfamiliar names, and the only thing you can understand is that somebody proved something. If this should happen to you, do not discard the book in anger. The purpose of a road map is to help you find your way in unfamiliar territory. If you were already familiar with the landscape, you would not need the map. The right way to use this book is to skip lightly over things you do not understand until you reach something that you are really determined to understand, and then turn to the bibliography. For students who seriously want to learn what modern mathematics is about, the most important part of the book is the bibliography. For casual readers who will not dig into the bibliography, for tourists in the land of mathematics who do not speak the local language, the road map still provides a wealth of useful information. The mathematical tourist can enjoy the human drama of John Fields and his medals without understanding the difference between a foliation and a functor, just as the spectator at the Winter Olympic Games can enjoy the triumph of Oksana Baiul in the figure-skating competition without understanding the difference between a triple lutz and a triple axel.

Preface

A famous physicist once gave this advice to authors publishing scientific papers. "Before submitting your paper for publication, hide it in a remote drawer of your desk, then take it out six months later and reread it. If it doesn't produce revulsion and a desire to throw it into the garbage, send it off."

This advice, like all good advice, has few adherents in real life, not even its own author. Circumstances not entirely within my control have caused this book to appear nearly six years after I wrote most of it. I thus could not take full advantage of the advice and adopted a compromise. The Russian edition was published in 1991. It contains mistakes made by the author, supplemented by many typographical and editorial lapses. As a compensation, the Russian edition includes an article by A.N. Kolmogorov "On tables of random numbers." This was an unusual event in publishing, in which a highly artificial combination of two authors resulted from the vagaries of the series "Mathematics and Cybernetics," from the publishing house Znanie, which printed my book.

However that may be, I was flattered by this co-authorship. After the passage of four years, the small annoyance experienced at the sight of misprints in this book has faded in comparison with the success of its publication in

Russian. In 1992 the series "Mathematics and Cybernetics," Znanie, and even the Soviet Union disappeared. And now the publishing of scientific books in Russia has become a dream. The English edition has undergone some revision in comparison with the Russian. Also added are several facts from the biography of John Charles Fields connected with the establishment of the prize. I gleaned them from the unpublished autobiography of John Lighton Synge, the Irish mathematician and physicist, who was a friend of Fields and a direct participant in this historic event.

Preface to the 1991 Russian Edition

This small book is an expanded version of an article published in *Istoriko-matematicheskie Issledovaniya* (*Historical Mathematical Research*) in 1989. The book was written later that year for Birkhäuser. The proposal by Znanie to publish it in the series "Mathematics and Cybernetics," in which I made my debut in 1979 with a brochure on Riemann, was, for me, a pleasant surprise. The compressed publication time meant that I had no chance to undertake an extensive revision of the text. In its main outlines, however, it does not seem to me to require any revision. The only serious addition is an analysis of the papers of the Fields medalists for 1990. The list of new Fields prize winners is in complete accord with the prediction that I made at the end of 1989.

The high level of Soviet mathematics was again confirmed at the 1990 mathematical congress in Kyoto. One

Fields medalist was the Soviet mathematician V.G. Drinfel'd. The Nevanlinna prize in applied mathematics was also awarded to a Soviet mathematician, A.A. Razborov. Both medalists were present at the awards ceremony. The invitation to give a plenary hour-long or a sectional 45-minute paper is an honor for any mathematician. Here the participation of Soviet mathematicians was significant. B.L. Feigin, G.A. Margulis, Ya.G. Sinai, and A.N. Varchenko gave hour-long talks. (There were only 15 hour-long talks altogether.) Eighteen people gave 45-minute sectional presentations. There were about 100 mathematicians in the Soviet delegation, and in contrast to previous years all invited speakers were free to participate in the congress. The absence of several speakers resulted from purely personal reasons.

Participants in the congress noted with satisfaction the propitious changes in the Soviet mathematical community. The changes largely arose from the improvement in the social climate of the USSR. Unfortunately the trying social situation in the country, the political instability, and the economic difficulties raise serious concerns for the future of science in general, and mathematics in particular.

Loss of the position of leadership won by many generations of Soviet mathematicians is not simply a loss in a single area of science. Mathematics is not only the foundation of all natural scientific knowledge, but it is an element of culture, in no way less important than music, literature, and art. In my view the forfeiture of the leading position in mathematics may cause irreparable damage to Russia's effort to occupy its rightful position of respect among developed nations.

Acknowledgements

For the publication of this book I am indebted to very many people through personal and written contact, especially L. Ahlfors, D. Anosov, M. Atiyah, A. Baker, F. Bogomolov, L. Boya, P. Cartier, A. Connes, P. Deligne, V. Drinfel'd, F. Dyson, G. Faltings, C. Ford, H. Furstenberg, K. Gawedzki, M. Gromov, A. Helemsky, G. Henkin, N. Hitchin, V. Kac, G. Margulis, D. Montgomery, S. Novikov, E. Primrose, K. Roth, B. Sanderson, G. Shabat, Y. Sinai, S. Smale, R. Thom, V. Vasil'ev, A. Wightman, and M. Yakobson. They and several other mathematicians have helped me to avoid serious errors and have expanded and improved my exposition.

Professors L. O'Raifertaigh and J. Lewis of the Institute of Advanced Studies in Dublin acquainted me with the autobiography of J.L. Synge. I would like to mention the interest and support of B. Zimmermann, who instigated the writing of this book and who unfortunately was not able to participate in its publication.

I worked on this book during visits to many research centers: Université de Paris-Sud (Orsay, France), Amsterdam University Institute of Theoretical Physics (Netherlands), Institut für Theoretische und Angewandte Physik (Stuttgart, Germany), Isaac Newton Institute for Mathematical Sciences (Cambridge, Great Britain), and Universität des Saarlandes. Professors M. Kléman, F.A. Bais, H.W. Capel, C.G. van Weert, E. Kröner, H.-R. Trebin, M.F. Atiyah, P. Goddard, T. Kibble, J. Wright, and A. Holz made these visits possible.

Finally, after many vicissitudes, this book is being published by A K Peters, Ltd. For preparing it for publication I am indebted to the enthusiasm of Klaus Peters and the translator Roger Cooke. The final editing of the translation was done by Ronald Calinger. I took my first steps in the publishing arena under the gentle direction of Klaus with my book *Riemann, Topology, Physics.* Fifteen years later we meet again, this time on his territory.

Michael Monastyrsky

Prologue

The author of a solely scientific book has no need to explain his reasons for writing it. When it comes to a work with an admixture of journalism, however, the situation is different. The author who values his scientific reputation must provide some explanation. In 1978 the Soviet mathematician G.A. Margulis was awarded the Fields medal along with three other leading mathematicians. This occasion was only the second on which this prize was awarded to a Soviet mathematician. On the first occasion, in 1970, the medal was awarded to S.P. Novikov. Both occasions should have been cause for joy in the Soviet mathematical community, as evidence of the international recognition of the high caliber of Soviet mathematical research. But, alas, this was not the case. The Soviet scientific community hardly learned what had occurred, and the scientific leadership disturbingly deprived both medalists of the chance to participate in the awards ceremony. The National Committee of Soviet mathematicians, which had long been headed by I.M. Vinogradov, made this decision. Vinogradov and, as unpleasant as it is to say, L.S. Pontryagin, were primarily behind the shameful act of excluding Margulis from the

Soviet delegation to the Helsinki Congress in 1978.[1]

At this time the idea came to me to write a small, rather popular note on the Fields prize, the history of its founding, and the works of the prize winners. It was originally planned to be published in the journal *Priroda* (*Nature*), a journal on the same level as *Scientific American* or *La Recherche.* The famous Soviet mathematician B.N. Delone had for many years been a member of the mathematical section of its editorial board. Delone was a legendary figure, who had in the past been a famous mountain climber, builder of sailplanes, artist, and musician. European-educated, he was a man of integrity and courage (and not only in mountain climbing). As evidence of this statement, he once journeyed to a prison camp in Mordovia with Academician A.D. Aleksandrov, where his grandson, the poet Vadim Delone, was being held after the invasion of Czechoslovakia in 1968. Additionally, for many years the *samizdat* writer Venedikt Erofeev, author of the controversial book *Moskva–Petushki*, lived at his dacha. Such activity is unusual in the context of the Soviet reality of those years. This small digression was made in order to prepare the reader for the ensuing course of events.

Originally Delone had been in favor of preparing this brief history, and the decision was made that we would write it together. But events took a dramatic turn. After the article was ready to go to press, Delone telephoned me to say that he had talked with Vinogradov and Pontryagin, who implored him not to print the article. Their

---

[1] The reader will find the original explanation of this decision in a letter of Academician Pontryagin to the journal *Science* (1979, Vol. 205, No. 4411).

influence was sufficient to prevent the article from appearing in *Priroda.* Shortly before his death in 1980, Delone regretted his momentary weakness, attributing it to the mesmerizing influence of Vinogradov, the evil genius of Soviet mathematics. In abbreviated form, under my name only, a partial note was published in the journal *Voprosy Istorii Estestvoznaniya i Tekhniki* (*Questions of the History of Science and Technology*), No. 2, pp. 72–75 in 1982.

The quite gloomy situation in Soviet mathematics in the 1970s led to the emigration of many talented young mathematicians, as well as some not so young. The situation improved slightly over the next few years, as new people came into positions of leadership following the death of Vinogradov in 1983. To a considerable degree the decisive reaction of the international scientific community prompted this improvement.

Ten years after the events described above, Birkhäuser unexpectedly proposed that I write a short book on the work of the Fields medalists. After some hesitation, I agreed. The reader can judge the work for himself. Let me add only a few remarks in passing.

1. Mathematics is a single subject, a fact that is not always obvious when you study the daily reality of research. It becomes clear, however, when you become acquainted with results obtained by great mathematicians. This realization is one by-product resulting from an analysis of the works of the Fields medalists. Although honors went to authors of the greatest achievements obtained in the year immediately preceding each congress and sometimes in areas of mathematics widely separated from one another, truly wonderful connections between them were discovered with the pas-

sage of time. For that reason an ε-grid over the works of the Fields medalists covers a significant portion of the achievements of modern mathematics.

2. The development of pure mathematics in the period between the two world wars, and especially in the post-World War II period, was characterized by weak connections with the applied sciences, in particular with physics. This association was especially true of the areas of mathematics in which many Fields medalists worked. It was difficult to imagine that the concepts of sheaf, étalé cohomology, J-functors, and the like would ever be applied in physics. It was still more difficult to imagine that physics could assist algebraic topology or geometry. This point of view was widespread. French mathematician Jean Dieudonné expressed himself unambiguously on this subject in 1962:[2]

> I would like to stress how little recent history has been willing to conform to the pious platitudes of the prophets of doom who regularly warn us of the dire consequences that mathematics is bound to incur by cutting itself off from applications to other sciences. I do not intend to say that close contact with other fields, such as theoretical physics, is not beneficial to all parties concerned; but it is perfectly clear that of all the

[2] In an address delivered at the University of Wisconsin in 1962 Dieudonné gave a survey of the achievements of the preceding decade in pure mathematics. He emphasized algebraic geometry, algebraic topology, complex analysis, and algebraic number theory. See J. Dieudonné, "The development of modern mathematics," *American Math. Monthly*, **71** (1964), 239–242.

> striking progress I have been talking about, not a single one, with the possible exception of distribution theory, had anything to do with physical applications.

But, as often happens with globally expressed opinions, the situation underwent a sea change 10 to 15 years later. Over the past few years a close union of modern physics and mathematics has developed that has been exceptionally productive for both sides. Moreover, the most esoteric areas of mathematics have found brilliant applications. At the same time, certain remarkable achievements in mathematics, which will be related in this book, are based on ideas that arose in papers on physics. A recent example is the solution of Schottky's problem, which makes use of the theory of nonlinear Kadomtsev-Petviashvili equations.

Prophecy is always dangerous, but one would like to think that the achievements of modern physical mathematics will be reflected in the names of the new Fields winners at the congress in Kyoto.

A Brief Biography of John Charles Fields

The founder of the prize, John Charles Fields, was born 14 May 1863 in Hamilton, Ontario, Canada. After graduating from the University of Toronto in 1884, he received his doctorate in 1887 from the Johns Hopkins University. In 1892 Fields traveled to Europe, where he mainly attended seminars in Berlin and Paris. During the following decade he became acquainted with such mathematicians as Ferdinand Georg Frobenius, Hermann Amandus Schwarz, and Lazarus Fuchs, and the physicist Max Planck. In 1902 he returned to Canada, to the University of Toronto, where he worked until the end of his life. Fields was a member of several academies, including the Royal Society of Canada (1907), the Royal Society of London (1913), and the Russian Academy of Sciences (1924).

In mathematics Fields stressed throughout his life the theory of algebraic functions and algebra, precisely the fields where, more than half a century later, the major achievements of many Fields medalists were concentrated. Fields himself provided an algebraic proof of the Riemann-Roch theorem, but his greatest renown resulted from his international activity. Through his efforts the International Congress was held in Canada in 1924.

Fields had to overcome the threat of a boycott of the congress, mostly by French mathematicians, if the Germans and their World War I allies were invited. Seemingly, tactical considerations related to this problem elicited the proposal from Fields to convene the congress in Toronto. He called it an "International Mathematical Congress." Other congresses had been called "International Congresses of Mathematicians."

Fields skillfully directed the organizational work of the congress and gave a paper on the theory of ideals. At this congress an international mathematical prize was first discussed. The final decision was not made until eight years later at the International Congress in Zürich.

While years of preparatory work had preceded the proposal for a Fields medal, the decisive events unfolded at the beginning of 1932. In January Fields wrote a memorandum entitled, "An international medal for outstanding achievement in mathematics." This memorandum discussed in detail the charter of the medal, the procedure for awarding it, and even general wishes as to the design of the medal. Fields was preparing to present his ideas in September at the congress in Zürich. But in May he fell gravely ill. Prof. John Lighton Synge, the secretary of the Toronto congress and close friend of Fields, recalled that a short time later he was suddenly called to Fields' side, whom he found in critical condition. A will was prepared, in which Fields gave most of his funds to the medal. According to Fields' will, Synge was to present the memorandum for adoption to the executive committee of the congress in Zürich.

Fields was not to learn of the decision. He died on 8 August 1932, one month before the congress opened.

The session of the ad hoc committee that considered the question of the prize was stormy. Not all the members of the committee supported the establishment of a prize. In particular Oswald Veblen spoke against it, perhaps motivated by the thesis that the study of science is its own reward, so the researcher has no need of additional encouragement. Nevertheless, most of the committee members favored Fields' proposal. At the plenary session of the congress, the question was finally decided in the affirmative.

There is an interesting story connected with the actual medal. As a Canadian patriot, Fields wanted the design made by a Canadian sculptor and proposed Tate MacKenzie, who was known for his memorials to war dead in Edinburgh, Cambridge (Britain), and Princeton. MacKenzie was inclined to depict the Greek mathematician Archimedes.

According to Fields' wishes, despite the stylistic discrepancy, the corresponding text was composed in Latin by Norwood, a Latin scholar. The Canadian mint cast the medal.

Since the first Fields medal was awarded at Oslo in 1936, 34 mathematicians have been so honored. The prize has become integral to the life of the international mathematical community.

The selfless activity of Fields has not been forgotten in Canada. The mathematical institute at the University of Waterloo, Ontario, which opened in 1992, is named in his honor.

History of the Fields Medals

On 21 August 1990 at the opening of the International Congress of Mathematicians in Kyoto, Japan, following tradition, the names of the Fields Medalists—the highest honors given by the International Mathematical Union in pure mathematics[3]—were announced. The Fields medal had passed its first big jubilee year—its fiftieth anniversary. Some estimate of the importance of this longevity can be gained from an excursion into the history of mathematical prizes.

In the nineteenth century prizes for outstanding scientific results were established in practically every European academy of sciences. Many were awarded to foreign scholars. But the mere fact of having been awarded did not make

[3] A prize in applied mathematics—the Nevanlinna prize—has also been given since 1982.

any of the awards permanent or significant throughout Europe. An ad hoc committee usually awarded each prize for the solution of a difficult problem, frequently through a competition. Either the prize ceased to be awarded later, or it was simply a one-time award. Among such prizes was that of Swedish King Oscar II, which went to Henri Poincaré.

The (Paris) Academy prizes were awarded more consistently, for example the Bordin Prize which was awarded to Sof'ya Kovalevskaya im 1888. However, like all the other mathematical prizes that existed in the nineteenth and early twentieth centuries, it had no claim to world- or even European-wide significance.

Alfred Nobel delivered the *coup de grâce* to mathematics by excluding it from the sciences designated to receive his prize.[4]

The role of prizes, like the role of international recognition in general, is important support for individual scholars. Despite Franz Neumann's beautiful quote: "The discovery of new truth is the greatest joy; recognition can add almost nothing to it," this wise idea is only partially true. Following Niels Bohr, the opposite conclusion is also valid. Recognition is especially important to young researchers.

The establishment of an international prize was first discussed in earnest in 1924 at the International Mathematical Congress in Toronto. This meeting was the second

[4] There exists a large number of conjectures, especially dear to the authors of popular articles "explaining" this fact. A critical analysis of this silly gossip is given in an article by the two Swedish mathematicians L. Hörmander and L. Gårding [HG].

international congress after World War I.[5] Mathematicians of different countries had healed the breaches caused by the war. It was natural to raise the question of establishing an international prize. Fields, the president of the congress, initiated this dialogue.

Establishing the prize was, however, far from simple. The world had entered upon a period of social cataclysms, and international cooperation was complicated by the interwar political map.

Fields' perseverance eventually yielded results, though not immediately. The next congress in Bologna (1928) reached no decision. However, by 1932, the year the subsequent congress was held in Zürich, several mathematical societies had reached a preliminary agreement. At the beginning of 1932 Fields wrote his memorandum which gave a detailed characterization of the charter for the new prize. This memorandum pointed out the basic properties that distinguished the new prize:

> One would here again emphasize the fact that the medals should be of such a character as purely international and impersonal as possible... There

[5] The first postwar congress met in Strasbourg in 1920. It was small in relation to the number of countries and participants, since no representatives of Germany or its allies were present. The congress in Toronto was more representative, although Weimar Germany and its allies were again not invited. A delegation from the USSR participated in this congress. Among the papers delivered by the Soviet mathematicians was the famous paper of B.N. Delone, "Sur les sphères vides." Unfortunately Delone himself did not take part in the work of the congress, and his paper was read by Prof. Y.V. Uspensky.

should not be attached to them in any way the name of any country, institution or person.

In contrast to the Nobel prize, there is no mention of Fields on the medal. The name of the winner and the year are engraved on the rim. Nevertheless, the name of Fields has deservedly become attached to both the prize and the medal.

The 1932 Zürich congress decided to give the first Fields medal at the next congress in Oslo in 1936. Fields' memorandum stated that the prize should not only recognize results already obtained, but also stimulate further research. The first Fields committee took this instruction to mean that the prizes should go to relatively young scholars.

The Oslo congress conferred the first Fields medals in 1936. The gold medals and the monetary prize of $1500 Canadian were awarded to Jesse Douglas (1897–1965) of the Massachusetts Institute of Technology for the solution of Plateau's problem and Lars Ahlfors (1907–1996) of the University of Helsinki for work on the theory of Riemann surfaces. More details on the work of Ahlfors and Douglas will be given below.

The choice of these first two medalists was quite important for setting standards. It established a certain age limit: all future medalists were under 40. In choosing candidates both the solution of difficult problems and the creation of new theories and methods enlarging the fields of application of mathematics were considered.

A special committee on the Fields medals, appointed by the executive committee of the International Mathematical Union, reviews candidates and selects winners. Usually, the chair of the Fields committee is the president of the union. The candidates are carefully identified. Many leading mathematicians are asked their opinions as a matter of course. The final selection is made by secret ballot through correspondence. The committee for this task changes after every congress. Names of the committee members, except for the chair, are kept secret until winners are announced at the next congress. These measures are designed to guarantee the maximum objectivity of the choice.

In 1936 the first Fields committee consisted of the leading mathematicians George D. Birkhoff, Constantin Carathéodory, Élie Cartan, Francesco Severi (chair), and Teiji Takagi. The next congress was scheduled to take place in the USA, four years later, but the Second World War upset that plan. Not until 1950, under the aegis of the reconstituted International Mathematical Union, was another congress called in Cambridge, Massachusetts, USA. It was held at Harvard University.

Soviet mathematicians had missed the 1936 congress. The two invited Soviet speakers, A.O. Gel'fond and A.Y. Khinchin, informed the organizing committee at the last minute that they could not attend. Removal of their papers from the congress program was announced at the opening ceremony. This act became a lamentable tradition at subsequent congresses.

Soviet mathematicians were also not allowed by their government to participate in the work of the 1950 congress. The campaign against "rootless cosmopolitans" was at its height—an antisemitic campaign that gradually became a struggle against all progressive trends in Soviet intellectual life. The acme of the "Lysenkovshchina,"[6] brought the attack on cybernetics, which resulted in the destruction of entire research specialties and schools in Russia. A spectacular example of the insular epistolary level of this era appears in a telegram from the president of the Academy of Sciences of the USSR, S.I. Vavilov, explaining why Soviet mathematicians could not participate in the work of the congress: "The Soviet Academy of Sciences thanks you for the cordial invitation to Soviet scholars to participate in the work of the international mathematical congress being held in Cambridge. Soviet mathematicians are too busy with their routine work and cannot attend the congress. I hope that the upcoming congress will be an important event for mathematical science. I wish you success in the work of the congress. (Signed) President of the Academy of Sciences of the USSR, S.I. Vavilov." A.N. Kolmogorov, one of the members of the Fields committee, thus could not participate in its work.

The Fields committee was enlarged to eight members

[6] The Russian suffix *-shchina* has no English equivalent and is attached to a name to denote a political campaign—usually sinister—associated with the person named. The Lysenkovshchina is an obscure phenomenon in Soviet biology. Lysenko and his supporters became notorious for abolishing the Soviet school of genetics. Lysenko's activity was dangerous because J. Stalin and N. Khrushchev supported it. In Soviet science the term "Lysenkovshchina" became a synonym for ignorance and arrogance in any field of science. The life of Lysenko and his teaching has been discussed in a paper by K.O. Rossianov, "Joseph Stalin and the 'new' Soviet biology," *Isis*, Vol. 84, No. 4 (1993), 728–795. See also the book of D. Joravsky, *The Lysenko Affair*, University of Chicago Press, 1986 [Jo].

in 1950, but met with only seven members. Through 1986, the committee consisted of eight members. In 1990 this tradition was broken when the membership of the committee was expanded to nine. Certain procedural changes were made in the composition of the 1994 committee. Vice-Chair David Mumford became chair of the committee, because of ethical considerations that arose from the nomination of Pierre-Louis Lions, son of the president of the International Mathematical Union Jacques-Louis Lions, for the award. V.I. Arnol'd, professor at the Université de Paris-Dauphine, also declined work on the committee for reasons of academic ethics.

A complete list of Fields committee members speaks more eloquently of their authoritativeness than any words. The place and country where the meeting was held is given in parentheses.

1950 (Cambridge, MA, USA): Harald Bohr (chair), Lars Ahlfors, Karel Borsuk, Maurice Fréchet, William Hodge, Damodar Kosambi, Andreĭ Nikolaevich Kolmogorov (did not participate), and Marston Morse.

1954 (Amsterdam, The Netherlands): Hermann Weyl (chair), Enrico Bompiani, Florent Bureau, Henri Cartan, Alexander Ostrowski, Arne Pleijel, Gabor Szegö, and Edward Charles Titchmarsh.

1958 (Edinburgh, Scotland): Heinz Hopf (chair), Komaravolu Chandrasekharan, Kurt Friedrichs, Philip Hall, Andreĭ Nikolaevich Kolmogorov, Laurent Schwartz, and Carl Ludwig Siegel.

1962 (Stockholm, Sweden): Rolf Nevanlinna (chair), Pavel Sergeevich Aleksandrov, Émil Artin, Shiing-Shen

Chern, Claude Chevalley, Lars Gårding, Hassler Whitney, and Kôsaku Yosida.

1966 (Moscow, USSR): Georges de Rham (chair), Harold Davenport, Max Deuring, William Feller, Mikhail Alekseevich Lavrent'ev, Jean-Pierre Serre, Donald Spencer, and René Thom.

1970 (Nice, France): Henri Cartan (chair), John Doob, Friedrich Hirzebruch, Lars Hörmander, Shokichi Iyanaga, John Milnor, Igor Rostislavovich Shafarevich, and Paul Turán.

1974 (Vancouver, B. C., Canada): Komaravolu Chandrasekharan (chair), John Adams, Kunihiko Kodaira, Bernard Malgrange, Andrzej Mostowski, Lev Semyenovich Pontryagin, John Tate, and Antoni Zygmund.

1978 (Helsinki, Finland): Deane Montgomery (chair), Ioan Mackenzie James, Lennart Carleson, Martin Eichler, Jürgen Moser, Yuriĭ Vasil'evich Prokhorov, Béla Szökefalvi-Nagy, and Jacques Tits.

1983[7] (Warsaw, Poland): Lennart Carleson (chair), Huzihiro Araki, Nikolaĭ Nikolaevich Bogolyubov, Paul Malliavin, David Mumford, Louis Nirenberg, Andrzej Schinzel, and Charles Terence Clegg Wall.

1986 (Berkeley, CA, USA): Jürgen Moser (chair), Michael Francis Atiyah, Pierre Deligne, Lars Hörmander, Kazufumi Ito, John Milnor, Sergeĭ Petrovich Novikov, and Conjeevaram S. Seshadri.

---

[7] This congress was scheduled for 1982, but the declaration of martial law in Poland in December 1981 caused it to be postponed to 1983.

1990 (Kyoto, Japan): Lyudvig Dmitrevich Faddeev (chair), Michael Francis Atiyah, Jean Michel Bismut, Enrico Bombieri, Charles Fefferman, Kenkichi Iwasawa, Peter D. Lax, Igor Rostislavovich Shafarevich, and John Griggs Thompson.

1994 (Zürich, Switzerland): David Mumford (Chair), Luis Caffarelli, Masoki Kashiwara, Barry Mazur, Alexander Schrijver, Dennis Sullivan, Jacques Tits, and S.R. Srinivasa Varadhan.

The prizes are awarded at the opening of the congress. After an introductory speech by the chair of the Fields committee the medals are presented by the honorary president of the congress. Among those conferring the medals have been the King of Sweden at the Stockholm congress in 1962 and the President of the Soviet Academy of Sciences Mstislav Vsevelodovich Keldysh at the Moscow congress in 1966.

In 1986 at the prize's fiftieth anniversary, Ahlfors, the first Fields medalist, was named honorary president of the congress. He shared his reminiscences of the first presenting of the prize. Names of the new medalists-to-be had been kept secret, so that Ahlfors learned only accidentally on the eve of the ceremony that he was the new winner and was officially notified only one hour before the congress opened. Such secrecy apparently resulted in Douglas, the second winner, not coming to the congress. The reason given was exhaustion from a long journey. But possibly more advance notice on the program might have given Douglas additional strength. Norbert Wiener accepted Douglas' medal on behalf of the Massachusetts Institute of Technology.

A special session at which papers devoted to the works of the winners are read precedes the scholarly program of the congress. Authorities in the corresponding fields of mathematics survey the achievements of Fields medalists. Although Carathéodory (1936), Bohr (1950), and Weyl (1954) presented synopses of all the prize-winning papers, subsequently a separate paper was dedicated to each winner.

Originally two prizes were established, but the improved fund and private contributions made it possible to give four prizes in 1966, 1970, 1978, and 1990 as well as three in 1983 and 1986.

The following mathematicians have been named Fields medalists.

1936: Jesse Douglas (1897–1965), Massachusetts Institute of Technology;[8] Lars Ahlfors (1907–1996), University of Helsinki.

1950: Laurent Schwartz (b. 1915), Université de Nancy; Atle Selberg (b. 1917), Institute for Advanced Study, Princeton.

1954: Jean-Pierre Serre (b. 1926), Université de Paris; Kunihiko Kodaira (b. 1915), Princeton University.

1958: Klaus Friedrich Roth (b. 1925), University of London; René Thom (b. 1923), Université de Strasbourg.

1962: Lars Hörmander (b. 1931), University of Stockholm; John Milnor (b. 1931), Princeton University.

1966: Stephen Smale (b. 1930), University of California, Berkeley; Paul Cohen (b. 1934), Stanford University;

---

[8] The institution named is the affiliation at the time the prize was awarded.

Alexander Grothendieck (b. 1928), Université de Paris; Michael Francis Atiyah (b. 1929), Oxford University.

1970: Alan Baker (b. 1939), Cambridge University; Sergeĭ Petrovich Novikov (b. 1938), Steklov Institute of the USSR Academy of Sciences; John Thompson (b. 1932), Cambridge University; Heisuke Hironaka (b. 1931), Harvard University.

1974: David Mumford (b. 1937), Harvard University; Enrico Bombieri (b. 1940), University of Pisa.

1978: Pierre Deligne (b. 1944), Institut des Hautes Études Scientifiques, Bures-sur-Yvette; Daniel Quillen (b. 1940), Massachusetts Institute of Technology; Grigoriĭ Aleksandrovich Margulis (b. 1946), Institute for Problems of Information Transmission, USSR Academy of Sciences; Charles Fefferman (b. 1949), Princeton University.

1983: Alain Connes (b. 1947) Université de Paris; William Thurston (b. 1946), Princeton University; Shing Tung Yau (b. 1949), Stanford University.

1986: Simon Kirwan Donaldson (b. 1957), Oxford University; Gerd Faltings (b. 1954), Princeton University; Michael Freedman (b. 1951), University of California at San Diego.

1990: Vladimir Gershonovich Drinfel'd (b. 1954), Physicotechnical Institute of Low Temperatures, Khar'kov; Edward Witten (b. 1951), Institute for Advanced Study, Princeton; Vaughan Jones (b. 1952), University of California at Berkeley; Shigefumi Mori (b. 1951), Kyoto University.

1994: Jean Bourgain (b. 1954), Institut des Hautes Études Scientifiques, Bures-sur-Yvette; Pierre-Louis Lions (b. 1956), Université de Paris-Dauphine; Jean-Christophe Yoccoz (b. 1957), Université de Paris-Sud, Orsay; Efim Zelmanov (b. 1955), University of Wisconsin, Madison/Institute of Mathematics, Novosibirsk.

Selection of young mathematicians supports the continuing development of mathematics. The Fields committees are representative of outstanding mathematicians of the older generation, which makes their assessment of the creativity of the young all the more interesting.

Mathematical Progress

Work of the Fields medalists encompasses nearly every branch of mathematics. Only the theory of probability has been passed over as of the present moment. Carrying out a detailed analysis of the medalist areas is tantamount to compiling an encyclopedia of modern mathematics. For this reason I shall only sketch the results obtained, singling out either those most amenable to brief exposition or closest to my own area of expertise.

An examination of the list of medalists shows that more than half of the prize winners work in algebraic topology, algebraic geometry, and complex analysis. This fact is quite revealing. Despite the largely cumulative and continuous development of mathematics in this century, the predominance of these features in the changing face of mathematics after World War II is indisputable. The three areas just listed are now so intertwined that it is difficult to separate them. It is even more difficult to place a boundary on the creativity of individual mathematicians, especially since the specialties of many have changed abruptly.

Topology

Jean-Pierre Serre. In 1954 Serre received the first Fields prize for a paper on topology. He is one of a brilliant constellation of French topologists, including such mathematicians as H. Cartan, and J. Leray. Only war and the age restriction prevented these others from becoming Fields medalists themselves. In particular Serre, a student of Cartan, applied the method of spectral sequences created by Leray to fundamentally advance the classical problem of topology—the computation of homotopy groups of spheres.

To assess Serre's contribution to the solution of this problem results of the preceding period must first be reviewed.

In 1935 Witold Hurewicz defined an n-dimensional homotopy group as the set of equivalence classes of homotopically distinct mappings of the n-dimensional sphere $\mathbb{S}^n$ into a topological space M^k. The standard notation for this group is $\pi_n(M^k)$. Hurewicz's definition is the natural n-dimensional generalization of the concept of the fundamental group $\pi_1(M^k)$. In 1895 H. Poincaré introduced the fundamental group, also called the Poincaré group, in his classical paper "Analysis Situs,"[9] which laid the foundation of modern topology.

In 1931 H. Hopf, in his paper "Über die Abbildungen der dreidimensionalen Sphäre auf die Kugelfläche," took a crucial step preceding the appearance of the general concept of a homotopy group. In modern terms he showed that the group $\pi_3(\mathbb{S}^2)$ is isomorphic to $\mathbb{Z}$, where $\mathbb{Z}$ is the

---

9 *Analysis situs*—analysis of position—is an archaic name for topology.

group of integers. Hopf's paper is a classic in topology. Its results and ideas stimulated the development of topology for many years. Among its new concepts are the Hopf invariant and the Hopf bundle.

Relatively recently, in the 1970s, it became clear that the Dirac monopole can be naturally interpreted in terms of the Hopf bundle. Curiously, P. Dirac's paper was also published in 1931, a few months after Hopf's paper appeared. Still, it took 40 years of development in physics and mathematics for the close connection between these papers to be noticed. Here is a worthy example for F. Dyson's collection of "missed opportunities" [Dy]. For more details on the connection of the Dirac monopole and the Hopf invariant see [Mo].

The nontriviality of the homotopy groups $\pi_n(M^k)$ for $n > k$ contrasts sharply with properties of homology groups $H_n(M^k)$ and cohomology groups $H^n(M^k)$, which are trivial for $n > k$. It is now possible to define homology and cohomology groups axiomatically, thereby guaranteeing a way of computing them. Knowledge of the homology and cohomology groups of a specific manifold is of very little help, however, in the study of the homotopy groups of the manifold. Hurewicz's isomorphism essentially establishes the only direct connection between the homotopy and homology of a given manifold:

Hurewicz' Theorem. $\pi_k(M) = H_k(M)$ *if* $\pi_i = H_i = 0$ *for* $0 < i \leq k-1$.

Consequently some profound pre-World War II results in the theory of homology and cohomology, including the creation of the foundations of the theory of characteristic classes, were of little help in computing homotopy groups.

In this area, the only general result was the suspension theorem proved by H. Freudenthal.

Freudenthal's Theorem. *The group $\pi_r(\mathbb{S}^n)$ is isomorphic to the group $\pi_{r+1}(\mathbb{S}^{n+1})$ for $1 \leq r < 2n-1$, and the group $\pi_{2n-1}(\mathbb{S}^n)$ is mapped onto the group $\pi_{2n}(\mathbb{S}^{n+1})$.*

Freudenthal succeeded in partially describing the kernel of the mapping:

$$f : \pi_{2n-1}(\mathbb{S}^n) \to \pi_{2n}(\mathbb{S}^{n+1}).$$

In particular he computed the group $\pi_4(\mathbb{S}^3) = \mathbb{Z}_2$.

Freudenthal's result is important. Using the Freudenthal isomorphism $\pi_r(\mathbb{S}^n) \simeq \pi_{r+1}(\mathbb{S}^{n+1}) \simeq \pi_{r+k}(\mathbb{S}^{n+k})$, one can show easily that the groups $\pi_{r+k}(\mathbb{S}^k)$ are independent of k for $k > r+1$. Taking $k = \infty$, we therefore obtain the so-called stable homotopy groups $\pi_r(\mathbb{S}^\infty)$. However, even for computing homotopy groups of type $\pi_{r+2}(\mathbb{S}^r)$, for example $\pi_5(\mathbb{S}^3)$, Freudenthal's theorem is useless. Thus each successive step in the computation of the homotopy groups requires a special technique that was achieved with great difficulty. Some methods invented to solve this problem were extraordinarily interesting in another circle of problems. Here first of all is the Pontryagin theory of framed manifolds. In 1950 Pontryagin studied submanifolds M^k of $\mathbb{R}^{n+k}$ that admit a field of n-linear independent normal vectors. In modern language this property is called trivializability of the normal bundle. The manifold M^k itself with the trivial frame is an n-framed submanifold in $\mathbb{R}^{n+k}$. Because of the property of trivializability, the manifold M^k admits a certain tubular neighborhood N diffeomorphic to $M^k \times \mathbb{R}^n$. It can be shown that there exists a mapping

$g : N \to \mathbb{R}^k$ such that $g\big|_{\partial N} \to \infty$. If $\mathbb{R}^{n+k}$ and $\mathbb{R}^n$ are regarded as the interiors of $\mathbb{S}^{n+k}$ and $\mathbb{S}^n$ respectively, the mapping g can be extended to a mapping $f : \mathbb{S}^{n+k} \to \mathbb{S}^n$ in such a way that the pre-image of a noncritical point $s_0 \subset \mathbb{S}^n$ is given by $f^{-1}(s_0) = M^k$.

Pontryagin made the crucial observation that the homotopy equivalence of mappings $f_1, f_2 : \mathbb{S}^{n+k} \to \mathbb{S}^k$ corresponds to a certain equivalence of framed submanifolds $M^k \subset \mathbb{R}^{n+k}$. This equivalence, called *intrinsic homology* by V.A. Rokhlin, amounts to the following: Two manifolds M_1^k and $M_2^k \subset \mathbb{R}^{n+k}$ are intrinsically homologous if a framed manifold (film) $W^{k+1} \subset \mathbb{R}^{n+k} \times R$ exists such that

1) $M_1 \times \{0\}$ and $M_2 \times \{1\}$ are obtained by intersecting W^{k+1} with the planes $x_{n+k+1} = 0$ and $x_{n+k+1} = 1$ respectively;

2) the normal bundles to the submanifolds M_1^k and M_2^k are the intersection of these hyperplanes with the normal bundle to W.

Following this route, Pontryagin computed the groups $\pi_{n+1}(\mathbb{S}^n)$ and $\pi_{n+2}(\mathbb{S}^n)$. Rokhlin, a colleague of Pontryagin's at the time, took the next, even more difficult step in using this technique to compute the group $\pi_{n+3}(\mathbb{S}^n)$. Further obstacles along this route proved technically insuperable. In 1954 the methods of the theory of framed manifolds became the basis for a promising new theory—cobordism theory.

The results of computations of the homotopy groups of spheres prior to Serre were often first-rate and very diverse. But general theorems and even approaches to general theorems were almost completely absent. Then, in a series of notes in the *Comptes Rendus*, and in more detail in his dissertation in the *Annals of Mathematics* (1951), Serre proved several general theorems on the structure of the groups $\pi_i(\mathbb{S}^n)$. Here is the statement of one of the most important.

The Serre Finiteness Theorem. *For even n the groups $\pi_i(\mathbb{S}^n)$ are finite for all $i > n$ except the group $\pi_{2n-1}(\mathbb{S}^n)$. The group $\pi_{2n-1}(\mathbb{S}^n)$ is the direct sum of $\mathbb{Z}$ and a finite group. For odd n and $i > n$ the group $\pi_i(\mathbb{S}^n)$ is finite.*

Interestingly, Serre based his proof on a connection that he found between the homotopy groups of $\mathbb{S}^k$ and certain homology and cohomology groups, not for the sphere itself, but for a certain associated space, the so-called loop space. The theory of spectral sequences can be applied in computing the homology and cohomology of the loop space. Following this route, Serre made progress in computing specific homotopy groups. He duplicated results of Pontryagin, J.H.C. Whitehead, and Rokhlin by computing $\pi_{n+1}(\mathbb{S}^n)$, $\pi_{n+2}(\mathbb{S}^n)$, and $\pi_{n+3}(\mathbb{S}^n)$, and in addition found the group $\pi_{n+4}(\mathbb{S}^n)$. No one had found this last result by other methods.

This general construction was applied with great success in computing the homotopy groups of other spaces, in particular symmetric spaces and Lie groups. The close connection between computing the homotopy groups of spheres and the classical Lie groups seems clear. As is known, the classical groups generate bundles with a spherical base. For example:

$$U(m+1) \xrightarrow{U(m)} \mathbb{S}^{2m+1}.$$

Serre solved a number of difficult problems connected with computing the homotopy groups of the classical Lie groups. In particular he successfully described the primary components of the groups $\pi_p(G)$ in many significant cases, showing that the primary components of the groups $\pi_q(G)$ are sums of the primary components of the groups $\pi_i(\mathbb{S}^j)$.

The problem of computing the homotopy groups of Lie groups turned out to be easier than the analogous problem for spheres. Yet even in the case of Lie groups specific problems connected with the structure of particular groups $\pi_i(G)$ remain unsolved. R. Bott proved a remarkable theorem on the structure of the stable homotopy groups of Lie groups. This theorem asserts that the stable groups $\pi_i(G)$ possess the property of periodicity, for example, $\pi_i(U(\infty)) = \pi_{i+2}(U(\infty))$.

Bott based his proof on a beautiful use of Morse theory. The periodicity theorem subsequently acquired fundamental importance in modern topology.

Although the achievements of Serre from the early 1950s are connected with topology, the theory of complex spaces had already entered his circle of interests. His interest seems to have been sparked by the influence of his teacher Cartan, the leading authority in the theory of complex manifolds. Cartan had worked in that area as early as the 1930s. It turned out that the Cartan-Oka theorems in Cousin's problem admit an effective statement in terms of cohomology with coefficients in analytic sheaves. The papers of Serre, partly written in collaboration with Cartan, greatly advanced the theory of analytic sheaves. The original definition of a sheaf by Leray in 1945 was supplemented by a number of definitions connected with delicate analyticity properties. Leray is credited with introducing two fundamental concepts that changed the face of postwar mathematics. It is a startling fact that he developed his principal ideas during the war, while in a German concentration camp. This period is mentioned in [Di1]. In particular, Serre studied the cohomology of complex spaces with coefficients in sheaves of holomorphic functions. Theorems on the structure of certain cohomology classes of analytic spaces entered the literature under the name of the Kodaira-Serre duality theorems. All these results were essential at the next stage of development of algebraic topology and geometry. Serre later turned to algebraic geometry and arithmetic, obtaining significant results in the theory of representations of p-adic groups, modular functions, and more. In addition he authored brilliant monographs and textbooks [S2, S3]. In 1986 Springer-Verlag issued the collected works of Serre in three volumes, giving a complete picture of his work [S1].

René Thom. One of the 1958 Fields medals was awarded to Thom, a representative of the same school of French topologists. Thom's papers constructed cobordism theory. One problem of this theory has a rather simple statement: find necessary and sufficient conditions for a given compact manifold M^n to be the boundary of a manifold W^{n+1}. A necessary condition—that the Stiefel-Whitney numbers be equal to zero—had been found earlier by Pontryagin. Thom proved the more difficult part, that this condition is also sufficient.

Thom's cobordism theory continued the work of the Soviet mathematicians Pontryagin and Rokhlin and gave rise to a number of outstanding papers that solved some very difficult topological problems. Later medalists further developed the ideas in Thom's papers.

In developing cobordism theory Thom introduced new topological concepts that have entered the basic lexicon of the modern topologist. One of these ideas—the concept of a Thom space—is key to the modern theory of characteristic classes. The Thom space is defined for a bundle of k-dimensional planes endowed with a Euclidean metric. Let ξ be a vector bundle over the manifold B, E the space of the bundle, and A the subset of $E(\xi)$ consisting of the vectors v with $|v| \geq 1$. Contracting A to a point, we obtain the space $E(\xi)/A$, which is called the Thom space $T(\xi)$. The space $T(\xi)$ can be identified with the one-point compactification of the space $E(\xi)$.

One remarkable application connected with the space $T(\xi)$ is a theorem that Thom proved on the connection of the homotopy groups of the spaces $T(\xi)$ and the homology groups of the manifold B, the base of the bundle of k-vectors.

Thom's Theorem. *The groups $\pi_{n+k}(T)$ are isomorphic to $H_n(B, \mathbb{Z})$ for all $n < k - 1$.*

This theorem and its generalizations form the basis for computations in cobordism theory, where the fundamental object became the group of cobordisms of manifolds. Thom remarked that the cobordism condition is an equivalence relation that makes it possible to introduce the concept of a group on the classes of cobordism manifolds, the cobordism group Ω. The group operation is taken as the disjoint sum of manifolds. The group of n-dimensional manifolds is usually denoted Ω_n. It is also natural to consider the group $\Omega = \Omega_0 + \Omega_1 + \cdots + \Omega_n \ldots$. The operation of forming the direct product $M^n \times N^m$ of two manifolds of dimensions m and n generates an additional operation:

$\Omega_n \times \Omega_m \to \Omega_{n+m}$, turning Ω into a (graded) ring. Some significant topological problems can be stated in terms of the structure of the ring Ω. Thom's predecessors Pontryagin and Rokhlin had obtained some results on the structure of the rings Ω_i, though not in such a general formulation. In particular, Rokhlin proved that the group Ω_3 is trivial, i.e., every oriented three-dimensional manifold is the boundary of a four-dimensional manifold. Thom described the group Ω_n using a relation he had found between the homotopy groups of the space $T(\xi)$ and Ω_n. Cobordism theory was subsequently developed in connection with the study of the groups Ω, taking account of additional manifold structure, for example the structure of complex manifolds, spin manifolds, and others. Computing each such group involves solving a challenging topological problem.

Many important ideas that have found applications, some many years later, are due to Thom. For example, Thom's use of Morse theory to study the topology of complex spaces was applied very recently by M. Goresky and R. Macpherson in analyzing the cohomology of manifolds with singularities [GM].

Of course, among Thom's main achievements are his papers on the theory of singularities, which led to its development into an independent discipline—catastrophe theory. Catastrophe theory, in somewhat vulgar terms, studies the global behavior of functions from properties of their singularities. This theory builds upon Morse theory, the Whitney theory of singularities, and much more. It consolidates numerous elegant results in different areas of mathematics in a unified conceptual framework. Both the proof of the fundamental theorems and, no less important, the statement of the problems, comes from Thom. Some solutions are fundamental. Here are the central ideas in this circle.

1. Given a mapping $f : \mathbb{R}^m \to \mathbb{R}^n$ that is structurally stable at a point $x_0 \in \mathbb{R}^m$, what does the normal form of such a mapping look like?

2. Give a description of the stable mappings $\mathbb{R}^m \to \mathbb{R}^n$.

Both of these problems require precise formulations of the concepts of stability, classes of mappings, and the like. It is not possible to go into the details here. The reader should consult the literature. The most complete mathematically rigorous exposition is a two-volume monograph [AGV]. Let me note just two important results of Thom relating to problems 1 and 2.

The first of these results is the classification of the stable mappings $\mathbb{R}^n \to \mathbb{R}^n$ for $n < 6$. All the normal forms can be written out. In higher dimensions the situation is more complicated. Thom showed that for $n \geq 9$ the stable mappings do not form a dense set in the class of all smooth (differentiable) mappings $\mathbb{R}^n \to \mathbb{R}^n$. Continuous invariants of the mappings f (moduli) exist. Therefore no substantive classification of the stable mappings in the class of smooth mappings exists. For topologically equivalent mappings, however, such a classification is possible. Thom's second result is:

There exists a topological classification of the germs of smooth mappings $f : \mathbb{R}^m \to \mathbb{R}^n$, if a set of infinite codimension is removed from the space of all mappings.

Thom stated this result in 1964, but the complete

proof was given only recently by A.N. Varchenko, a student of V.I. Arnol'd.

In the context of singularity theory it became possible to explain many seemingly mysterious and accidental connections between the results obtained in branches of mathematics far removed from one another. It turned out, for example, that the classification of degenerate critical points of functions is determined by the Dynkin diagrams of semi-simple Lie algebras.

Numerous leading mathematicians are working productively today in singularity theory. They include the large school in Moscow founded by Arnol'd, as well as E.C. Zeeman and his students in Warwick, England. Catastrophe theory has gone beyond pure mathematics and even physics. Ideas of this theory are being applied in such diverse areas as economics and sociology. Next to Fermat's last theorem, catastrophe theory has become the mathematical phenomenon best known to the public. Thom himself was led first by ideas of his theory to study biological systems and later linguistics. His book *Stabilité Structurelle et Morphogénèse* (1972) aroused great interest among specialists and has been translated into several languages [Th].

John Milnor. Milnor, who had obtained important results in computing the cobordism groups of manifolds, received one of the 1962 prizes. In particular, he solved the problem that Thom had left open on the orders of the torsion subgroups Γ_k of the groups Ω_n for $n \geq 8$. Milnor, and independently of him the Moscow mathematician B.G. Averbukh, proved that the subgroups Γ_k have no elements of odd order. This result requires use of powerful topological techniques: the Steenrod operation and the Adams spectral sequence. C.T.C. Wall later succeeded in describing the group Ω_n completely by using the results of Milnor and Averbukh. Another result in cobordism theory obtained by Milnor and independently by S.P. Novikov was the computation of the unitary cobordism group, i.e., the ring of cobordism manifolds with a unitary structure group. This class defines the complex cobordism and corresponds to manifolds with a quasi-complex structure.

Milnor's most brilliant result, however, was the proof in 1956 that there exist different smooth structures on the seven-dimensional sphere. This discovery, which caught the imagination of all mathematicians, led to the creation of a new area of topology—differential topology.

Milnor based his original proof on the introduction of an invariant of the differentiable structure of a manifold. Consider a simply connected compact manifold M^7 with cohomology groups $H^3(M^7, \mathbb{Z}) = H^4(M^7, \mathbb{Z}) = 0$. The Milnor invariant $\lambda(M^7)$ is constructed as follows. Using Thom's result that $\Omega_7 = 0$, one can choose a manifold W^8 whose boundary is M^7. On the manifold W^8 the following topological invariants are defined: the first Pontryagin class p_1 and the signature $\sigma(W^8) \in H^4(W^8, \mathbb{Z})$. From these invariants the quantity λ is determined as a function of p_1 and σ.

Milnor primarily observed that λ depends only on the choice of the cobordism class of the manifold W^8, and not on W^8 itself. Using earlier computations by F. Hirzebruch of the signature $\sigma(W^8)$, Milnor obtained the following expression for λ:

$$\lambda = 45\sigma + p_1^2\langle W^8\rangle = 7p_2\langle W^8\rangle.$$

Here p_2 is the second Pontryagin class and $\langle W^8 \rangle$ is the fundamental homology class of W^8.

All that remained was to construct manifolds M_i^7 such that $M_i^7 = \partial \widetilde{W}^8$. The manifold $\widetilde{W}^8$ is a bundle over the sphere $\mathbb{S}^4$ with fiber equal to the ball B^4, whose boundary is the sphere $\mathbb{S}^3$. Thus the manifolds M_i^7 form a bundle of three-dimensional spheres over $\mathbb{S}^4$. The signature $\sigma(\widetilde{W}^8)$ and $p_1(\widetilde{W}^8)$ were known. For $\lambda(M_i^7)$ one obtains:

$$\lambda(M_i^7) = \lambda(\widetilde{W}^8) = 45 + p_1^2 + 7p_2.$$

It is known that $p_1 = k$ for the disk bundle $\widetilde{W}^8$, where k is any integer congruent to 2 modulo 4. We have $p_2 = (45 + k^2)/7$. But if $k \neq \pm 2 \pmod 7$, then the number p_2 is not an integer. Since the Pontryagin numbers of smooth manifolds are integers, and the manifolds M_i^7 are homeomorphic to the standard sphere $\mathbb{S}^7$, we obtain Milnor's result.

Milnor and the Swiss mathematician M. Kervaire succeeded in describing all the different smooth structures on $\mathbb{S}^7$. They demonstrated that a group operation can be introduced on the set of smooth structures $\theta(\mathbb{S}^7) = \theta_7$ that makes θ_7 into an abelian group. The group θ_7 is the finite cyclic group of order 28, $\mathbb{Z}_{28}$.

The sphere $\mathbb{S}^7$ is not the only example of a manifold with different smooth structures. Milnor and Kervaire proved that the group $\theta(\mathbb{S}^n)$ is finite for $n > 7$ and computed it in some cases, for example $|\theta(\mathbb{S}^{11})| = 992$. Quite recently the existence of a large number of smooth structures on $\mathbb{S}^{11}$ has been analyzed in the context of the membrane theory of supergravity, where the manifold $\mathbb{S}^{11}$ arises in compactifying the additional degrees of freedom of the space-time continuum.

The study of smooth structures on manifolds has proceeded in many directions. Kervaire constructed an example of a ten-dimensional manifold having no smooth structure [Ke].

Milnor's 1956 proof that 28 different smooth structures exist on the seven-dimensional sphere is by no means a pathological result. Shortly afterward the German mathematician E. Brieskorn constructed all 28 smooth structures on $\mathbb{S}^7$, defining them by the following systems of equations in $\mathbb{C}^5$:

$$z_0^3 + z_1^{6k-1} + z_2^2 + z_3^2 + z_4^2 = 0, \quad k = 1, \ldots, 28;$$
$$|z_0|^2 + |z_1|^2 + |z_2|^2 + |z_3|^2 + |z_4|^2 = 1.$$

Here z_0, $z_1, \ldots$, z_4 are arbitrary complex numbers.

S. Donaldson showed 25 years later that there exist different smooth structures on a compact simply-connected four-manifold. This outstanding result gained him the Fields medal in 1986.

Milnor is a versatile mathematician who has made outstanding contributions in the theory of discrete and algebraic groups and K theory. In recent years Milnor has been studying a topic connected with rational mappings of complex domains. This field of research, begun in the early twentieth century in the papers of French mathematicians P. Fatou, G. Julia, P. Montel, and others, is developing rapidly at present. Some important and unexpected connections are now being discovered with ergodic theory, quasi-conformal mappings, discrete groups, and fractals.

Milnor's beautiful books on Morse theory, characteristic classes, and cobordisms [Mi1, Mi2] expound complicated branches of mathematics with exceptional clarity.

After Milnor three other topologists were added to the ranks of the Fields medalists: M.F. Atiyah, S. Smale, and S.P. Novikov.

Michael Francis Atiyah. Atiyah, a 1966 medalist, is the author of significant results in several branches of algebraic topology and complex analysis. His first papers on algebraic surfaces put him into the first ranks of mathematicians. But his chief result was the index theorem, proved in 1963 in collaboration with American mathematician I. Singer.

The index theorem for elliptic operators on an arbitrary compact manifold M^n can be stated as follows.

Let D be an elliptic differential operator on M^n. It is known that the kernel of the operator D (ker D) forms a finite-dimensional vector space. Similarly the concept of the cokernel can be defined as coker $D = \ker D^*$ (here D^* is the adjoint operator). The quantity $\dim(\ker D - \ker D^*)$, called the *index* $(\operatorname{ind}(D))$ turns out not to change under continuous deformations of the manifold M^n, i.e., it is a topological invariant. I.M. Gel'fand conjectured that $\operatorname{ind}(D)$ can be expressed in terms of the characteristic classes of the manifold M^n, and Atiyah and Singer proved that such an expression is possible.

The index theorem has long roots. It has subsumed many classical results connecting the topological properties of manifolds with their differential-geometric properties. For example, the simplest case of the index theorem is a theorem of Poincaré expressing the sum of the indices of a vector field on a surface in terms of the Euler characteristic. Here the operator D is $\partial/\partial\bar{z}$. This theorem was subsequently generalized. First, the class of manifolds was enlarged, for example, the index theorem was extended to manifolds with boundary, open manifolds, and so forth. Second, the class of operators was enlarged. In essence, a larger class of operators was used in the original proof—pseudodifferential operators, whose theory had been developed in the preceding years. The 1962 Fields medalist Lars Hörmander made a major contribution to the construction of the general theory of pseudodifferential operators.

The index theorem has also been applied in the theory of complex algebraic varieties. A special case of it is the Riemann-Roch-Hirzebruch theorem—a key result in algebraic geometry.

The studies of Atiyah and Singer on the index problem began with attempts to generalize the Riemann-Roch-Hirzebruch theorem. While doing this work they learned of Gel'fand's conjecture from Smale. Details and other subtle remarks of Atiyah appear in his interview in the *Mathematical Intelligencer* [At].

The original proof of the Atiyah-Singer theorem was complicated. It used a wide spectrum of mathematical concepts, from the methods of topological K-theory, where the fundamental results were due to Atiyah, Hirzebruch, and others, to cobordism theory, to the theory of pseudodifferential operators, Sobolev spaces, and subtle facts from functional analysis.

The proceedings of several yearly seminars and workshops devoted to the study of the index theorem, which appeared soon after Atiyah and Singer's 1963 paper, give an idea of the difficulty of the first proof. The significance of this theorem has not diminished, even today. In recent years interesting applications have been found in physics. Quantum anomalies in field theory and computing dimensions of the spaces of instantons in gauge theories remain issues. The index of the Dirac operator and its analogues are also computed in these theories.

Application of the index theorem in physics helped simplify the proof itself. After the first proof there were several attempts to simplify and clarify it. A greater simplicity and lucidity marks the proof of the authors [At] and the papers of Atiyah, Bott, and V.K. Patodi [At], but these works relied on the same body of ideas, even though they eliminated one complicated ingredient or another.

Physicist E. Witten took a new approach to the proof of the index theorem. He originally applied the ideas of supersymmetry, which originated in the 1970s, to prove the Morse inequalities and Lefschetz's formula. It became clear from the conceptual framework of the index theorem that this approach could be applied to the theorem itself. Physicist L. Alvarez-Gaumé [AG] obtained this result.

Here, briefly, are the main ideas of the proof of Alvarez-Gaumé. Choose some one-dimensional quantum-mechanical supersymmetric system. In the Euclidean formulation of field theory this is a system in the space of $(0+1)$-dimensions with Hamiltonian H and a system of supercharge operators (supersymmetric charges) Q^i satisfying the system of commutation relations

$$\{Q^i, Q^{j*}\} = 2\delta^{ij} H, \quad \{Q^i, Q^j\} = \{Q^{i*}, Q^{j*}\} = 0, \qquad (1)$$

where $\{Q^i, Q^j\} = Q_i Q_j + Q_j Q_i$ is the anticommutator.

In the $(0+1)$-dimensional field theory one of the coordinates can be interpreted as time, and the usual concept of spin does not exist. Hence it is not obvious how to define an operator that maps boson states into fermion states. E. Witten [Wi1] introduced the fermion operator $(-1)^F = \exp(2\pi J_z)$, which makes it possible to map bosons into fermions. Here J_z is a projection defined in the Hilbert space of states.

We now introduce the operator $S = \frac{Q+Q^*}{2}$. It follows from (1) that

$$S^2 = H. \tag{2}$$

Witten observed that the eigenstates of the Hamiltonian H with nonzero energy map into other states of nonzero energy under the operator S, but these states have opposite fermion numbers $|E\rangle$. I use the standard Dirac notation from the physics literature. Therefore all states with nonzero energy arise in the form of Fermi-Bose pairs that generate a two-dimensional representation of the supersymmetry for each energy level. For the zero values of the energy this is no longer the case. Therefore the number of boson states with zero energy $n_B^{E=0}$ is not equal to the number of fermion states with zero energy $n_F^{E=0}$. Witten showed that the difference $n_B^{E=0} - n_F^{E=0} = W$ is $\mathrm{Tr}\,(-1)^F$; more precisely that it equals the regularized trace

$$\mathrm{Tr}\,(-1)^F e^{-\beta H} = n_B^{E=0} - n_F^{E=0}. \tag{3}$$

Little remains to be done in order to see that Eq. (3) defines the index of the operator Q on the Hilbert space of states. When the Hilbert space of states is decomposed into boson and fermion parts the condition $Q|\Psi\rangle = 0$ defines this decomposition. Thus one can define the index of the operator $Q = \mathrm{ind}\,(Q) = \dim\,(\ker Q - \ker Q^*)$; and as

Witten showed,

$$\mathrm{Tr}(-1)^F = \dim(\ker Q - \ker Q^*). \tag{4}$$

The index of the operator Q can now be computed using the technique of functional integration that was well developed in the literature. Using perturbation theory, we compute the integral

$$\mathrm{Tr}\,(-1)^F e^{-\beta H} = \int d\varphi(t)\, d\psi(t)\, \exp(-S_E(\varphi, \psi)), \tag{5}$$

where $S_E(\varphi, \psi)$ is the Euclidean action. Next expand the quantity $\mathrm{Tr}\,(-1)^F e^{-\beta H}$ into a series in β. Then the first term in (5), which is independent of β, gives the index of the operator Q.

The scheme for computing the index of any elliptic operator Q thus consists of the following. Construct a suitable one-dimensional supersymmetric quantum-mechanical model with supercharge Q. For this model determine the Hilbert space of states in which the operator Q acts. The index of the operator Q equals the Witten index of the model.

Alvarez-Gaumé [AG] showed that to find the indices of such classical operators as the Euler characteristic, the Hirzebruch signature, and the index of the Dirac operator on a compact manifold, it suffices to consider a supersymmetric extension of an elementary σ-model on M^n

defined[10] by the Lagrangian L_F (free motion of a point over M^n):

$$L_F = \frac{1}{2} g_{ij}(\varphi)\dot{\varphi}^i \dot{\varphi}^j,$$

$$L_{\text{sup}} = \frac{i}{2} g_{ij}(\varphi)\overline{\Psi}^i \gamma^0 D_t \psi^j + R_{ijkl}\overline{\Psi}^i \Psi^k \overline{\Psi}^j \Psi^l.$$

Here

$$D_t \Psi^i = \frac{d\Psi^i}{dt} + \Gamma^i_{jk}\dot{\varphi}^j \Psi^k; \ \ \overline{\Psi}^i_\alpha = \Psi^i_\beta \gamma^0_{\beta\alpha}, \ \ \alpha, \beta = 1, 2;$$

$\gamma^0 = \sigma_2$ (the Pauli matrix); and $\Psi^i = \begin{pmatrix} \Psi^i_1 \\ \Psi^i_2 \end{pmatrix}$ is a two-component real spinor.

The full Lagrangian has the form $L = L_F + L_{\text{sup}}$.

By going into more depth in the topic of the index theorem I have attempted to illustrate how modern physical ideas have facilitated a fresh look at fundamental mathematical achievements. The interweaving of old and new physical and mathematical ideas and methods produces an exceptionally strong impression.

---

[10] The σ-model, so popular in physics in recent years, is nothing more than harmonic mappings. Its study in mathematics began in the late 1950s and early 1960s in papers of F. Fuller, J. Eells, J. Sampson, and others. However, the influence of physical ideas, the supersymmetry of the σ-model, the quantum σ-model, and the like have considerably increased interest in the theory of harmonic mappings in recent years. This quiet, respectable area of mathematics was peripheral to the problems of most interest in the 1970s.

The result of Atiyah and Singer gave rise to a new branch of mathematics—global analysis. In a happy phrase of Atiyah, topologists used to study simple operators on complicated manifolds while analysts studied complicated operators on simple spaces. The time has arrived to study complicated operators on complicated spaces.

Undoubtedly, the numerous applications of the Atiyah-Singer theorem in different branches of mathematics and physics will ensure it a long life. In recent years Atiyah, without breaking his connection with pure mathematics, has engaged in studies of modern physical mathematics. Here also he obtained several outstanding results. It suffices to mention his paper on the classification of instantons, written with V.G. Drinfel'd, N. Hitchin, and Yu.I. Manin, and his paper on the representation of four-dimensional instantons as loop spaces over two-dimensional instantons. Together with his friends and co-authors Bott and Singer, he has developed the Yang-Mills theory over Riemann surfaces.

Atiyah is the author of numerous survey and popular articles on modern problems of mathematics and mathematical physics, and he is an excellent lecturer and teacher. Not surprisingly, a student of Atiyah, Donaldson, made a remarkable discovery in four-dimensional topology using ideas of field theory and algebraic topology. Oxford University Press has published the complete works of Atiyah up to 1985 in five weighty volumes (*in folio*) [At]; but they no longer give a complete picture of Atiyah's works, since his indefatigable research leads yearly to the publication of several new and substantive papers.

Stephen Smale. In deepening our acquaintance with the prize-winning papers on topology, we turn to those of

Smale, the 1966 winner. Of his remarkable results in the topology and theory of dynamical systems, let us begin with topology.

The Poincaré conjecture is among the most difficult problems of topology. In modern terms it can be stated as follows:

Poincaré Conjecture: *A closed smooth simply connected manifold M^n with the homology groups of the sphere $\mathbb{S}^n$ is homeomorphic to $\mathbb{S}^n$.*

Poincaré stated this conjecture in three dimensions. The natural generalization to the n-dimensional case is called the generalized Poincaré conjecture. Poincaré believed that a stronger assertion was true, namely that M^n is diffeomorphic to $\mathbb{S}^n$. But, as follows from the existence of Milnor's exotic spheres, the conjecture is not true in this form. Smale proved a more general theorem on h-cobordism, from which it follows that the Poincaré conjecture holds for dimensions $n \geq 5$. In dimensions five and six the stronger Poincaré conjecture is true (M^n is diffeomorphic to $\mathbb{S}^n$).

Smale's theorem on h-cobordism is stated as follows: Let V, V', and W be a triple of manifolds, and let V and V' be boundaries of W; assume that V and V' are deformation retracts of W. This statement defines h-cobordism.

Smale's Theorem. *If V and V' are simply connected, then $W \sim V \times I$, and consequently V is diffeomorphic to V'.*

Smale's proof, which skillfully uses the theory of Morse surgery, is beautifully described in Milnor's book [Mi2].

At first sight it seems paradoxical that the proof of the Poincaré conjecture for higher-dimensional spaces is more accessible than for three- and four-dimensional manifolds. The reason is that a map of a surface into a manifold of fewer than five dimensions cannot be approximated by an embedding. The situation is similar to the classification of manifolds.

M. Freedman and Donaldson, the 1986 prize winners, have made the most significant progress in recent years in four-dimensional topology. We shall turn to their works shortly.

Lacking space to review all of Smale's achievements in topology, I shall mention only one beautiful result that has a curious geometric interpretation. In 1959 Smale proved that any two immersions of $\mathbb{S}^2$ into $\mathbb{R}^3$ are regularly homotopic. One corollary of this theorem is that a sphere can be turned inside out in $\mathbb{R}^3$. This result and its generalization have attracted attention recently in connection with the study of groups of diffeomorphisms of spheres.

Smale wrote another series of papers on the theory of dynamical systems, which have a profound topological basis. The application of topology in the theory of dynamical systems is long standing. Among Smale's predecessors are George D. Birkhoff, who studied in detail the behavior of three-dimensional systems in a neighborhood of a homoclinic trajectory. Poincaré discovered homoclinic trajectories in his famous study of the restricted three-body problem in celestial mechanics. In the 1920s and 1930s Birkhoff examined dynamical systems having phase manifolds of complicated structure. A prolonged period of mostly specialized investigations followed. Smale deserves credit for reviving interest in the theory of dynamical systems and the construction of a multi-dimensional theory, which over the last 30 years has become one of the most prolific areas of mathematics. One important concept of the multi-dimensional theory of dynamical systems is provided by the so-called structurally stable systems. For this class Smale constructed a substantive theory. In 1937 A.A. Andronov and Pontryagin had precisely stated the concept of structural stability of a dynamical system, but they studied it only for the two-dimensional case (in the plane).

The concept of stability of a system of differential equations can be stated as follows. A system of equations

$$\dot{x} = f(x), \tag{6}$$

where x is an n-dimensional vector, is called *structurally stable* if the topological type of the trajectories (phase portrait) is preserved under a small perturbation in the right-hand side. A precise formulation of structural stability requires the introduction of several concepts (cf. [Sm1]).

In the two-dimensional case structurally stable Andronov-Pontryagin systems have these properties:

a) a finite number of singularities of focus and saddle type;

b) a finite number of limit cycles.

Smale showed that in the multi-dimensional case the situation alters radically. He constructed a structurally stable system having an infinite number of singular points, limit cycles, and so forth. Smale's book [Sm2] gives interesting details connected with this discovery. He showed that structurally stable systems also arise from discrete automorphisms, for example, the group of automorphisms of the two-dimensional torus generated by the transformations with eigenvalues λ_1, λ_2 such that $\lambda_1 = 1/\lambda_2$, $\lambda_2 > 1$. Smale conjectured that the geodesic flow on manifolds of negative curvature is structurally stable. This conjecture, later proved by D.V. Anosov, led to the identification of an important class of dynamical systems with exponentially unstable trajectories ($\mathcal{Y}$-systems or Anosov systems).

Currently the theory of multi-dimensional dynamical systems is finding physical applications in turbulence and hydrodynamics and has been enriched by such notable discoveries as strange attractors and Feigenbaum universality.

In conclusion we make one remark of historical character. Apparently the first example of a strange attractor is the Lorenz attractor. In 1963 E. Lorenz made a numerical study of the following system of equations, which describes a trimodal convection equation. This system is a good approximate description of the motion of the atmosphere:

$$\dot{x} = ax + by$$
$$\dot{y} = xy + bx - y$$
$$\dot{z} = xy - cz$$

(a, b, and c are fixed numbers). An invariant set was found for this system that behaved like trajectories of Smale type. A remarkable, but exceptionally difficult, 1945 paper of the British mathematicians M. Cartwright and J. Littlewood studied the Van der Pol equation with a perturbation of the form

$$\ddot{x} - k(1 - x^2)\dot{x} + x = b\mu k \cos(\mu t + \alpha).$$

They observed Smale behavior in the trajectories in a certain domain of the parameter b (infinitely many periodic solutions, unstable trajectories, etc.).

Of course, neither Cartwright nor Littlewood could have foreseen the prolific development of the theory of dynamical systems. They both simply exchanged their usual area of work for one more suited to wartime. For Cartwright, a specialist in differential equations, this change was more natural. But Littlewood also laid aside his beloved theory of numbers and, with his characteristic analytic brilliance, took up the solution of a difficult and, as it must have seemed to them, applied problem. After all, the Van der Pol equation arises in radiophysics. Littlewood apparently was seriously interested in this problem, for he returned to the Van der Pol equation in 1957 and devoted two long memoirs to it.

Sergei Petrovich Novikov. In 1970 Novikov was honored with the Fields medal. He is the author of outstanding results throughout topology. This account begins with his unique work on the theory of foliations, which lies at the interface of dynamical systems and topology. The theory of foliations is a multi-dimensional generalization of the theory of ordinary differential equations in the following sense. Instead of trajectories, the theory of foliations considers distributions of hypersurfaces defined by differential forms. The simplest class of foliations is the class of foliations of codimension one (or the class of $n-1$-dimensional hypersurfaces in an n-dimensional manifold). The theory of foliations is a relatively new branch of mathematics located at the juncture of differential topology and the theory of differential equations. It essentially begins with a paper from the 1940s by C. Ehresmann and G. Reeb, who constructed a nontrivial foliation on $\mathbb{S}^3$. This foliation was not smooth. In 1952 Reeb improved the construction slightly, obtaining a smooth foliation that now bears his name.

The Reeb Foliation. The sphere $\mathbb{S}^3$ is represented as $|z_1|^2 + |z_2|^2 = 1$, $z^i = (\rho^i, \theta^i)$, and the Reeb foliation is constructed first on the anchor ring $D^1 \times \mathbb{S}^1$, $|z^1|^2 \leq 1/2$, where D^1 is the disk with coordinates ρ^1, θ^1, and $\mathbb{S}^1$ is the unit circle parametrized by the angle θ^2. The foliation is defined geometrically using the following sections:

a) $\theta^2 = c$, $(\rho^1)^2 = \text{const}$ (the section is a circle), $(\rho^1)^2 = 0$ (the section is a point);

b) a section of the foliation $\tan\theta^1 = \lambda$ consists of the curves $\theta^2 = \theta_0^2 = \exp[-1/(1 - 2(\rho^1)^2]$ for $\rho^1 < \sqrt{2}/2$. Foliations on the anchor ring $\rho^2 < \sqrt{2}/2$ are constructed similarly. The border $\rho^2 = \rho^1$ forms a closed fiber. This fiber is the only one of the Reeb foliation; all other fibers are homeomorphic to the plane.

The three-dimensional sphere $\mathbb{S}^3$ can be obtained by gluing the two anchor rings $\rho^2 < \sqrt{2}/2$, and $\rho^1 < \sqrt{2}/2$ together, identifying the points of the boundary $\rho^2 = \rho^1$. Choosing one Reeb foliation on each polytope and taking account of the identification of the boundaries, we obtain the Reeb foliation on all of $\mathbb{S}^3$. Here a compact fiber of $\mathbb{S}^3$ is T^2. This foliation has smoothness C^1.

According to one conjecture, any foliation of codimension 1 on $\mathbb{S}^3$ has a compact fiber. Novikov proved this difficult conjecture. In the proof he set forth constructions that were important for the development of the theory of foliations.

For codimension two (curves) this conjecture is false. The corresponding conjecture made in the 1930s by H. Seifert was refuted in 1974 by P. Schweitzer, who constructed an example of a vector field of smoothness C^1 on $\mathbb{S}^3$ for which there is no periodic solution, i.e., no compact fiber of codimension two.

Novikov wrote other papers involving algebraic topology. One classifies the simply connected manifolds M^n for $n \geq 5$ (a result obtained independently by W. Browder); another paper proves the topological invariance of the rational Pontryagin classes. Although Novikov's theorem on the topological invariance of the Pontryagin classes is true only for simply connected manifolds, its proof involves passing to nonsimply connected toroidal submanifolds. Despite some artificiality of this device, attempts to derive a new proof have as yet come to nothing.[11] In 1966 Novikov advanced a conjecture on the structure of the homotopy invariants of nonsimply connected manifolds, which came to be known as the conjecture on higher signatures. It asserts that for manifolds with nontrivial fundamental group ($\pi_1(M^n) \neq 0$) all homotopy invariants can be represented as integrals of the Hirzebruch polynomials $L(p_1, \ldots, p_n)$ multiplied by classes defined by π_1, where $p_i(M^n)$ are the Pontryagin classes. This conjecture greatly interested topologists. The efforts of many mathematicians went into the proof of Novikov's conjecture for a large class of manifolds. G. Lustig proposed a new method to prove the Novikov conjecture on higher signatures based on the theory of elliptic operators, a gener-

[11] The short analytic proof by D. Sullivan and N. Telemann of Novikov's result still uses Novikov's reduction at a key point [ST].

alization of the Atiyah-Singer theory. The development of Lustig's method, through some modifications from mathematical physics, enabled A. Connes, H. Moscovici, and J. Lott to prove the conjecture for manifolds for which $\pi_1(M^n)$ is a hyperbolic group. This class of groups, introduced by M. Gromov, is particularly important for the study of three-dimensional manifolds in the context of Thurston's program.

In later years Novikov turned away from algebraic and differential topology and immersed himself in mathematical physics, where he obtained important results in the general theory of relativity—the structure of homogeneous models—and nonlinear integrable systems. One fundamental result in this direction was the solution of the periodic problem for the Korteweg-de Vries equation. This equation was studied independently by P. Lax. The corresponding formulation of the problem led to the creation of a new area in the theory of integrable systems closely connected with algebraic geometry. Considering the two-dimensional generalizations of the Korteweg-de Vries equations, the Kadomtsev-Petviashvili equations, Novikov arrived at an intriguing conjecture that is connected with the solution of a classical problem of the theory of Riemann surfaces—Schottky's problem. This problem consists of finding a system of equations that characterizes the Jacobian of a Riemann surface. Novikov proposed that the corresponding theta functions describe the Jacobian if and only if the theta function is a solution of the Kadomtsev-Petviashvili. This Novikov conjecture was recently proved by T. Shiota and M. Mulase.

Through close collaboration with physicists Novikov pursued the study of gauge fields, where he pioneered results in the theory of multivalued functionals. Here, as in his other physical papers, Novikov combines physical intuition with profound topological technique. Together with his students, he has recently obtained important results in string theory.

His close contact with physicists moved Novikov to discuss putting modern geometry and topology into a form accessible to theoretical physicists. Many physicists and mathematicians had recognized the need for such an exposition, but Novikov was the first to carry out this plan, together with his colleagues B.A. Dubrovin and A.T. Fomenko. Study of his three-volume *Modern Geometry* gives the theoretical physicist the necessary grounding for work in new branches of mathematical physics.

An analysis of the research of topologists shows that many of them are connected with the problem of the classification of manifolds. This section on topology closes with a discussion of papers devoted to three- and four-dimensional manifolds.

Michael Freedman. In 1982 Freedman, who was a 1986 prize winner, proved the Poincaré conjecture for $\mathbb{S}^4$. His result—the classification of four-dimensional compact simply connected topological manifolds M^4—was more general than the conjecture itself. Taking into account the intersection form Q that arises on the cohomology group $H^2(M^4, \mathbb{Z})$ his result is: *any unimodular quadratic form over $\mathbb{Z}$ (the ring of integers) can be an intersection form Q on $H^2(M^4, \mathbb{Z})$.*

As often happens in mathematics, the simple statement of the result conceals an extremely difficult proof.

The main tool was a complicated surgery technique that involves gluing on so-called Casson handles. The idea of using the intersection form for classification goes back to a 1952 paper of Rokhlin, who obtained a result that was not fully appreciated until 20 years later. He showed that if a smooth compact simply connected manifold has an even intersection form Q, its signature σ must be divisible by 16, while for topological manifolds it suffices that σ be divisible by 8. As Donaldson later proved, this result means that manifolds M^4 with an even definite form Q cannot have a smooth structure.

In recent years Freedman's interests have become attached to physics. But in contrast with Donaldson's work, which was initiated by the theory of gauge fields, Freedman sought to apply topology to plasma physics and magnetohydrodynamics. He succeeded in estimating the energy of dissipation of a magnetic field using the nontriviality of the linking of magnetic lines of force. [FH, FHW]

Simon Donaldson. Only a year after the 1982 publication of Freedman's result, Donaldson, who was a 1986 prize winner, proved that for a smooth four-dimensional compact simply connected manifold M^4 with a positive-definite intersection form Q, this form can be diagonalized over the ring $\mathbb{Z}$, i.e., $Q = x_1^2 + \cdots + x_n^2$. In conjunction with Freedman's result, this proof could lead to an unexpected corollary—there exist homeomorphic but not diffeomorphic four-dimensional manifolds.

Donaldson's paper does not imply that different smooth structures on $\mathbb{S}^4$ exist, and thus does not refute the strong Poincaré conjecture. For $\mathbb{S}^4$ we have $H^2(\mathbb{S}^4, \mathbb{Z}) = 0$, and the theorem becomes trivially degenerate. Nevertheless a possible strengthening of Donaldson's results leads to the construction of different smooth structures on $\mathbb{S}^4$. Donaldson's results imply that there exist different smooth structures on the noncompact space $\mathbb{R}^4$ (four-dimensional Euclidean space).

In a series of papers R. Gompf and C. Taubes constructed first a countable (Gompf) and then a continuous (Taubes) family of pairwise nondiffeomorphic smooth structures. Nevertheless the family constructed by Taubes is incomplete, since there are exotic $\mathbb{R}^4$ spaces (fake $\mathbb{R}^4$) that do not occur in the Taubes series. The properties of such spaces are unusual. For example, compact subsets exist in them that cannot be surrounded by a three-dimensional sphere imbedded in the standard manner. The occurrence of exotic $\mathbb{R}^4$ spaces may have significant implications for quantum field theory, since the differential structure of Euclidean space is at issue.

Donaldson's results generated a new field of research that revealed unexpected properties of manifolds. Particularly interesting are those connected with the differential structure of algebraic varieties.

The question arises: what are the relations between the topological and algebraic properties of four-dimensional manifolds? The answer can be given in terms of the matrix of intersection indices, or, the form Q. Donaldson's result was unexpected. He constructed an entire series of algebraic surfaces which have the same quadratic form Q but which are differentiably inequivalent. This result is connected with another property of such manifolds. For topological manifolds Freedman proved the property of decomposability, i.e., if the form Q satisfies $Q = Q_1 + Q_2$, then a manifold M with form Q can be represented as $M_1 + M_2$

with forms Q_1 and Q_2 respectively. For the algebraic surfaces described by Donaldson this property does not hold. Such algebraic surfaces are said to be *indecomposable.* Several beautiful conjectures relating to the structure of indecomposable manifolds exist. Atiyah proposed the following: The indecomposable smooth four-dimensional manifolds are the sphere $\mathbb{S}^4$ and the indecomposable algebraic surfaces. In the summer of 1990 Gompf and T. Mrówka refuted Atiyah's conjecture when they constructed an example of a surface of type $K3$—a simply connected manifold of dimension four that is not diffeomorphic to any algebraic (complex) surface.

I shall mention a more interesting result in this body of ideas. S.M. Finashin, M. Kreck, and O.Ya. Viro [FKV] have proved the following theorem: *There exists an infinite series $S_1, S_2, \ldots$ of smooth two-dimensional surfaces of $\mathbb{S}^4$ such that*

1) *for any i and j the pairs $(\mathbb{S}^4, S_i)$ and $(\mathbb{S}^4, S_j)$ are homeomorphic, but not diffeomorphic;*

2) *each S_n is homeomorphic to the connected sum of ten copies of the real projective plane RP^2.*

Donaldson's papers are full of interesting results and methods of proof. Along with a subtle topological technique, he employed constructions of modern field theory—instanton solutions of the Yang-Mills equations. Here the important papers of Taubes and K. Uhlenbeck form the basis of the analytic part of Donaldson's proof. In their books D. Freed and Uhlenbeck [FU], and H.B. Lawson, Jr. [L2] gave a beautiful exposition of Donaldson's theorem, which includes all the necessary analytic and topological machinery.

Since there is not space to mention all of Donaldson's achievements, I shall list just two results from the classification of monopoles and instantons. Both of these problems come from physics, but their complete solution became possible only by use of the techniques of algebraic geometry. For the problem connected with the classification of instantons, Donaldson found a new method of proof. The second problem is linked with the complicated problem of describing the moduli space of monopoles (three-dimensional classical time-dependent solutions of the field equations). Interesting unsolved problems on the connection of monopole solutions with the classification of three-dimensional manifolds arise here, by analogy with the four-dimensional case.

In 1985 A. Casson of the California Institute of Technology introduced a new class of integer invariants of three-dimensional manifolds M^3 (mostly homological spheres). His approach is related to representations of the group $\pi_1(M^3)$ into a gauge group, for example $SU(2)$ acting in connection with zero curvature over M^3. Casson's result was given an important generalization by A. Floer,[12] who constructed the homology group of infinite-dimensional spaces of connections. Here also a connection with physical models can be traced. This area remains in its infancy, but connections already have been noted with symplectic Morse theory, pseudoholomorphic curves (in the sense of Gromov) in symplectic manifolds, and invariants of four-dimensional manifolds augur substantial and unpredictable results.

---

[12] The German mathematician Floer (b. 1956) died in 1991 at the beginning of a brilliant research career.

William Thurston. The early 1930s saw the completion of a classification of two-dimensional manifolds that is remarkable for its simplicity and completeness. But the transition to three-dimensional spaces posed enormous difficulties that seemed insuperable. This situation changed radically after the papers of Thurston, a 1983 prize winner. Ideas of Thurston and a few other mathematicians promoted progress in tackling even this seemingly hopeless problem.

Thurston conjectured that a three-dimensional manifold admits a canonical decomposition into pieces having a certain geometric structure. This last phrase needs decoding. It asserts that a compact three-dimensional orientable manifold can be partitioned into pieces by two-dimensional spheres and tori imbedded in it in such a way that by gluing the boundary spheres of three-dimensional balls and leaving the tori alone we obtain a manifold M^3 with boundary admitting a geometric structure. The manifold M^3 is said to admit a geometric structure if it is possible to introduce on it a complete locally homogeneous Riemannian metric. In the most interesting cases we are talking about a metric of negative curvature (hyperbolic spaces).

Thurston's conjecture has not been proved completely. A proof of it would contain a proof of the three-dimensional Poincaré conjecture. However, for a large class of manifolds, the so-called Haken manifolds, Thurston was able to prove his conjecture. Thurston also conjectured that any hyperbolic manifold can be covered by a Haken manifold.

The connection of the topology of three-dimensional manifolds with differential-geometric characteristics (the existence of a hyperbolic metric on a manifold) seems unexpected. A number of important results have been obtained in this direction. For example, Gromov proved that for hyperbolic spaces the volume is a topological invariant; moreover there exists only a finite number of complete hyperbolic spaces having volume less than a given constant.

From the study of three-dimensional manifolds beautiful connections have been discovered between different branches of mathematics—the theory of Kleinian groups, quasi-conformal mappings, discrete groups, dynamical systems, and several others. These connections stand out especially clearly in the problem of describing up to isotopy the group of diffeomorphisms (or homeomorphisms) $\{\varphi\}$ of two-dimensional surfaces M^2. Among those studying this problem in the 1930s O. Teichmüller obtained the chief results. Teichmüller approached the problem from the point of view of the theory of Riemann surfaces and quasi-conformal mappings, and J. Nielsen studied the problem from the geometric and algebraic points of view. Nielsen called attention to the importance of considering the partition of M^2 into pieces even in the case of closed surfaces M^2 and studying the properties of the mappings φ under approach to the boundary. Thurston remarkably used the theory of dynamical systems in this problem and, in particular, introduced a class of foliations on M^2 that are hyperbolic and Anosov unstable. Here it is necessary to introduce a hyperbolic metric in the space of surfaces of genus larger than one. The reader drawn to this promising field of mathematics should consult the popular article [TW] and the survey of Thurston [T2]. Unfortunately, as of this writing Thurston's 1979 Princeton lectures exist only as a preprint.

We close this brief excursion into the papers of Thurston by noting the many brilliant auxiliary results in his studies of three-dimensional manifolds. Thurston, for example, proved a long-standing conjecture of P.A. Smith.

Theorem (Smith's Conjecture). *Let $\varphi : \mathbb{S}^3 \to \mathbb{S}^3$ be a diffeomorphism that preserves orientation, let $\varphi^n = 1$, and suppose there exists a fixed point of the mapping φ. Then the set of fixed points of φ is an unknotted circle and the diffeomorphism φ is conjugate to an isometry.*

Thurston found rich results in other areas of mathematics, for example the theory of foliations. Here is one example: on each manifold with $\chi(M) = 0$ there exists a foliation of codimension 1. Recently Thurston has studied models of cellular automata.

The Thurston papers reflect a growing interest in the study of the subtle connections between the algebraic-geometric, differential-geometric, and topological characteristics of manifolds.

Complex Analysis

Shing Tung Yau. At this point it is natural to build a bridge to work in complex analysis. Violating the chronology, I begin with the papers of 1983 medalist Yau, a mathematician of wide ranging interests. One of Yau's achievements was the proof of a 1954 conjecture of Calabi.

Calabi's Conjecture. *Let M^n be a compact Kähler manifold with the Kähler metric g and associated form Ω:*

$$\Omega = (i/2)g(z, \bar{z})\, dz^i \wedge d\bar{z}^j . \tag{7}$$

Every Kähler form (7) *is cohomologous to the form* $\widetilde{\Omega}$ *generated by the Ricci tensor.*

Numerous corollaries of this theorem attest to its importance for complex analysis and algebraic geometry. We shall mention one of them, known as the Severi conjecture: *If a complex surface is homotopic to* CP^2 *(the two-dimensional complex projective plane), then it is biholomorphic to* CP^2.

Currently, the study of manifolds with a Ricci-flat metric (the first Chern class $c_1 = 0$), known as Calabi-Yau manifolds, has increased in connection with new discoveries in physics, notably in string theory and superstrings. In string theory physical space-time is a 26-dimensional space, the so-called *bosonic string*, and a ten-dimensional space, known as the *fermionic string*. Complicated problems arise in the attempt to pass to four-dimensional space-time by compactifying the extra degrees of freedom. The physical requirements lead to the assumption that the compactified manifolds are Calabi-Yau manifolds. These manifolds include also those with singularities, or orbifolds in the terminology of Thurston.

Another result of Yau having a significance that reaches beyond mathematics was the proof with R. Schoen of the conjecture that mass is positive in the general theory of relativity. They proved that for a nontrivial isolated physical system the total energy, including the contribution of matter and gravitation, is positive. Later Witten advanced a new proof using ideas of supersymmetry.

Omitting discussion of other results of Yau, I note his papers on three-dimensional manifolds, co-authored with W. Meeks. They solved some long-standing problems in the theory of minimal surfaces. The results thus obtained, such as the equivariant loop theorem, were important for proving the Smith conjecture also. In a series of papers coauthored with Lawson, Yau described a class of manifolds having a metric of positive scalar curvature acted on by a compact nonabelian group of transformations.

Although the proofs proposed by Yau often include complicated analytic machinery (for example, a priori estimates of solutions of the Monge-Ampère equation are used to prove the Calabi conjecture), the results themselves are algebraic or topological in nature.

Lars Ahlfors. Let us go back 50 years. For his work in complex analysis, Ahlfors received the first Fields medal. He is the representative of a brilliant school of Finnish mathematicians founded by E. Lindelöf and R. Nevanlinna. Ahlfors helped found the modern geometric theory of Riemann surfaces. In his famous 1935 paper "Zur Theorie der Überlagerungsflächen," (Acta Math., **65**, 157–194), he exhibited the class of surfaces for which the Nevanlinna theorems on the distribution of values of meromorphic functions hold. He constructed a theory of coverings and demonstrated that the corresponding surfaces are determined by a wider class of mappings than conformal mappings. Ahlfors called these mappings *quasi-conformal.* H. Grötzsch had introduced these mappings somewhat earlier in 1928. Also in 1935 M.A. Lavrent'ev introduced them in the theory of quasi-elliptic equations. Largely due to Ahlfors himself, the importance of quasi-conformal mappings in the theory of complex manifolds was soon realized. The machinery of quasi-conformal mappings makes it possible to solve basic problems in the theory of Riemann surfaces: the description of the space of moduli of Riemann surfaces and the Teichmüller space, the deformation of Riemann surfaces, and so on. Ahlfors deserves credit for reviving interest in the fundamental papers of Teichmüller during the 1930s. In their work Ahlfors and his students, especially Lipman Bers, strengthened the results of Teichmüller and developed them in several directions. Especially beautiful are their studies of Kleinian groups. For over half a century, Ahlfors has been the leader in complex analysis, confirming the soundness of his selection by the first Fields committee. His collected works were published in 1982 [Ah].

Kunihiko Kodaira. At the 1954 congress in Amsterdam the Fields medal was given to Kodaira. His papers encompass three branches of mathematics: topology, complex analysis, and algebraic geometry. The artificiality of separating these branches is easily seen through Kodaira's papers.

A most important Kodaira paper concerns a criterion for a compact complex manifold to be algebraic. He proved that a compact complex manifold is algebraic if and only if it is a Hodge manifold, i.e., it has a Kähler form cohomologous to an integral form. Kodaira also provided the first multi-dimensional generalizations of the Riemann-Roch theorem and the first classification of compact complex surfaces.

In 1954 Kodaira published the fundamental paper "On Kähler varieties of restricted type" [Kod], which contained the proofs of important theorems on the degeneracy (triviality) of certain cohomology groups of compact complex manifolds. His theorems (vanishing theorems) are crucial for investigations into the geometry of complex manifolds. One corollary of Kodaira's theorem is a simple proof of

Lefschetz's theorem connecting the cohomology of a projective manifold with the cohomology of a nondegenerate hyperplane section of it.

Kodaira's papers were the crowning achievement of a long period of research into algebraic manifolds over the field of complex numbers. In the opinion of the distinguished mathematician Hermann Weyl, these papers were the foremost contribution to the theory of complex manifolds since those of Hodge.

Weyl, who chaired the Fields committee in 1954, delivered a speech on the papers of the medalists Kodaira and Serre. Curiously, Weyl had difficulty distinguishing the areas of research of the two mathematicians. He said, "The uninitiated may get the impression that our Committee erred in awarding the Fields Medals to two men whose research runs on such closely neighboring lines. It is the task of the Committee to show that, despite some overlap in methods, they give the solutions of completely different, extremely difficult problems."

Continuing his predilection for the theory of complex manifolds, Kodaira later obtained several important results, particularly in his series of joint papers with D. Spencer on the deformation of analytic spaces. These papers investigated families of complex varieties. They opened new areas of investigation. The works of Kodaira up to 1975 have been published in three volumes by Princeton University, where he worked for nearly 30 years. Since that time the connections between algebraic geometry, topology, and complex analysis have become even closer.

Algebraic Geometry

Alexander Grothendieck. In 1966 Grothendieck received a Fields medal for work in algebraic geometry. The name of Grothendieck is linked with a revolution in algebraic geometry, which influenced other areas of mathematics. The concept of schemes that he introduced raised algebraic geometry to a new level of abstraction, beyond the reach of mathematicians with a traditional education. The theory of sheaves, spectral sequences, and other innovations in the late 1940s and early 1950s are subsumed by this complicated technique. But if certain mathematicians could console themselves for a time with the hope that all this complicated structure was "abstract nonsense,"[13] the later papers of Grothendieck and his successors showed that classical problems of algebraic geometry and the theory of numbers, the solutions of which had resisted efforts of several generations of talented mathematicians, could be solved in terms of the Grothendieck K-functor, motives, l-adic cohomology, and other equally complicated concepts.

The outstanding results obtained in this body of ideas garnered prizes in subsequent years. Grothendieck, who was at the summit of his fame in the early 1970s, has faded from the mathematical horizon. He has not published a single article in the past 20 years. Nevertheless, one of his manuscripts that has become available to mathematicians shows the depth and intensity of his work during this time. The manuscript "Esquisse d'un Programme," [14] presented

---

[13] In algebra the term "abstract nonsense" has a definite meaning without any pejorative connotation.

[14] A book analyzing this work has recently been published by the Cambridge University Press [Grot1].

to the CNRS[15] to apply for a research position at the University of Montpellier, contains a detailed exposition of several topics that Grothendieck believes to be important for the development of mathematics, and which he wanted to study with colleagues and students. In essence it is a question of finding connections between geometry, combinatorial topology, and algebraic geometry. One problem that Grothendieck considers can be stated as follows: Consider a certain graph $\mathcal{D}$ (which Grothendieck calls a *dessin*) on a Riemann surface M^2 satisfying certain regularity conditions; in particular it is a connected one-dimensional complex. To each such graph one can assign a smooth algebraic curve $X_{\mathcal{D}}$ defined over some number field. The existence of such a curve is a corollary of a complicated theorem proved by Moscow mathematician G. Belyi.

The central question is the following: Suppose a graph $\mathcal{D}$ is given. What can be said about $X_{\mathcal{D}}$ and functions $\beta_{\mathcal{D}} : \mathcal{D} \to X_{\mathcal{D}}$? In the simplest case, when $\mathcal{D}$ is a tree on the sphere $\mathbb{S}^2$, the corresponding curves are hyperelliptic functions.

The converse statement of the problem is also of interest. Given an algebraic curve, to which graphs on the surface does it correspond? Grothendieck claimed that the correspondence between the graphs and the functions $\beta_{\mathcal{D}}$ is determined by the action of the Galois group Gal $(\overline{\mathbb{Q}}/\mathbb{Q})$ on the graph $\mathcal{D}$. (Here $\mathbb{Q}$ is the field of rational numbers and $\overline{\mathbb{Q}}$ is its algebraic closure.) The connection between geometric properties of graphs on a surface and algebraic functions discovered by Grothendieck is of great value not only from

[15] Centre Nationale de Recherche Scientifique, the organization that funds most of the basic research in France.

the purely mathematical point of view, but also for theoretical physics, especially in such branches as string theory and crystallography. One possible application is the theory of two-dimensional gravitation; another is the theory of two-dimensional and three-dimensional quasi-crystals. Moscow mathematicians V. Voevodskii and G. Shabat, trying to prove some of Grothendieck's conjectures, have obtained elegant results by studying graphs on a surface of genus ≤ 2. They confirm Grothendieck's conjecture and raise the curtain on some difficult general theorems.

I have mentioned only one of the themes contained in Grothendieck's "Esquisse d'un Programme." Undoubtedly a careful study of other propositions and conjectures will lead to no less interesting discoveries.

Since I cannot refer the reader directly to Grothendieck's "Esquisse d'un programme," I recommend reading the three-volume festschrift dedicated to Grothendieck's sixtieth birthday [Grot2]. It gives a rather complete picture of the influence of Grothendieck on modern mathematics.

Heisuki Hironaka. In 1970 Hironaka was awarded the medal for solving an important problem—the resolution of the singularities of algebraic varieties over fields of characteristic zero.

A variety with singularities is defined as a variety for which the Jacobian fails to have maximal rank at some points. The simplest example of a variety with singularities is provided by curves of "node" and "cusp" types. Typical examples of such curves are $y^2 = x^3 + x^2$ (node) and $y^2 = x^3$ (cusp). Another series of varieties with singularities is obtained by the following construction. Consider a manifold M^n on which a discrete group Γ acts with fixed points $x_i \in M^n$. Then the space M^n/Γ is a variety with singularities. Such varieties, as already stated, are now called *orbifolds.* Some classes of orbifolds, for example $M^6 = \mathbb{T}^6/\Gamma$, where $\mathbb{T}^6$ is the six-dimensional torus, occur in modern string theory.

The problem of resolving singularities in algebraic geometry consists of the following. Given an algebraic variety $\widetilde{M}$ with singularities, is it possible to construct a nonsingular variety M such that $\widetilde{M}$ can be obtained through a birational mapping of M? In algebraic geometry a process of resolution of singularities (the σ-process) has been developed, in which a projective plane of suitable dimension is glued in place of the singular point. In this way the process of successive resolution of singularities arises. When this procedure is done, however, many complicated problems arise; for example, one must prove that the removal of a singularity does not cause new singularities to appear.

The problem of resolution of singularities is regarded as central in algebraic geometry. Resolution of singularities for curves was known even in the nineteenth century, while for surfaces it appeared in the works of the Italian school of algebraic geometry. But, like many other results of this school, it was considered nonrigorous, since it was obtained by transcendental methods. At present almost all of the results of the Italian school have been rehabilitated. In 1939 O. Zariski gave a purely algebraic proof over fields of characteristic zero. Finally, in 1964 Hironaka solved this difficult problem for n-dimensional algebraic varieties in characteristic zero.

David Mumford. In 1974 Mumford received the prize. He had greatly advanced the solution of a classical problem of algebraic geometry, namely the description of the moduli spaces of abelian varieties. Mumford's papers include a purely algebraic construction of the theory of theta-functions.

His papers on invariant theory revived interest in this classical area of mathematics, connected with the names of D. Hilbert, P. Gordan, A. Clebsch, and other remarkable late nineteenth and early twentieth century mathematicians. Mumford completely transformed invariant theory from the point of view of modern algebraic geometry. In particular he introduced the concept of stability of vector bundles. This concept originally arose in describing the orbits of a group G on an algebraic variety. Later it became clear that in those cases when the algebraic variety is a parameter space or a family of algebraic objects, and the group G establishes an equivalence between them, the stability of the orbit of an object is faithfully reflected in its geometric properties. This observation lay at the basis of the constructive theory of moduli of algebraic varieties, and made it possible to solve many specific problems of algebraic geometry by the method of degeneracy. The essence of this approach can be illustrated by the following example.

Consider the space of solutions of a polynomial equation $f(z) = 0$, where $f(z)$ is a polynomial of degree n in the complex variable z. The number of roots of this equation equals n not only for a polynomial in general position, when all the roots are distinct, but also for any polynomial of degree n if the roots are counted with the appropriate multiplicities. But degeneracies can also be bad, for example, if the coefficient of z^n becomes zero.

Mumford's theory constructively distinguished "good" stable degeneracies in almost any specific geometric situation, leading eventually to the solution of many classical problems of algebraic geometry. For example, Mumford and J. Harris proved that the moduli space of algebraic curves of large genus is nonrational, and Harris wrote on the connectedness of manifolds of smooth curves with elementary modal singularities.

Mumford's papers are at the forefront of promising investigations of recent decades, directed toward merging the generalizing concepts of modern algebraic geometry and the brilliant particular results of classical mathematics obtained by transcendental methods, in particular by the Italian school of algebraic geometry.

Pierre Deligne. In 1978 Deligne received the prize for a proof of a conjecture of André Weil on zeta-functions over finite fields. In 1949 Weil had stated a series of conjectures on the behavior of the analogue of the Riemann zeta-function for algebraic varieties over fields of finite characteristic. His conjectures included the proof of the rationality of the zeta-function and the behavior of the "zeros" (α_i) of the zeta-function, which is the analogue of the famous, still-unproved Riemann hypothesis for the classical zeta-function: $\operatorname{Re}\alpha_i = 1/2$.

For curves over finite fields, E. Artin, H. Hasse (for elliptic curves), and Weil himself had stated and proved these results, which are great achievements in algebraic number theory. The study of zeta-functions of curves over finite fields stimulated the development of the powerful machinery of algebraic geometry and made it possible to solve some difficult problems in number theory. Among other things, the Italian school of algebraic geometry found purely algebraic proofs for a number of classical theorems. Despite individual successes, however, (for example B. Dwork's proof of the rationality of the zeta-function), the multi-dimensional Weil hypotheses remained inaccessible. Deligne's proof in 1973 is striking in its beauty and complexity, but required the application of the wealth of techniques accumulated in algebraic geometry over the preceding years. Here the work of M. Artin and Grothendieck was decisive, in particular the concept of l-adic cohomologies introduced by Grothendieck. The construction of the l-adic cohomologies allowed the extension of the fundamental results of Lefschetz for cohomologies of classical algebraic varieties (over the field $\mathbb{C}$) to varieties over finite fields. Nicholas Katz [Kat] gave a beautiful and maximally accessible exposition of Deligne's theorem. The proof of several classical hypotheses of number theory follows from Deligne's result.

The classical Ramanujan conjecture is a particular consequence of Deligne's results. This conjecture can be interpreted as a statement about the behavior of the coefficients of the cusp form Δ. In 1930 the German mathematician H. Petersson offered some interesting conjectures on the behavior of the coefficients of more general modular forms. Afterwards this body of ideas became known as the Ramanujan-Petersson conjectures. Deligne succeeded in proving the Ramanujan-Petersson conjectures in general form.

Ramanujan Conjecture. *Consider the parabolic form*

$$(2\pi)^{-12}\Delta(z) = x\prod_{n=1}^{\infty}(1-x^n)^{24} =$$

$$= \sum_{n=1}^{\infty} \tau_n x^n, \quad x = \exp(2\pi i z).$$

Then $|\tau_p| \leq 2p^{\frac{11}{2}}$ *for all primes* p.

This statement is quite simple, but a proof aimed at a mathematician who is not a specialist in algebraic geometry, in the words of Deligne, would occupy about 2000 pages.

Despite the diligent efforts of many mathematicians, including Deligne, who later devised a new proof of the Weil conjecture, it has not been possible to simplify the proof in any important way. The most complicated part of the proof is using l-adic cohomologies. Nevertheless Gerard Laumon has obtained certain fundamental simplifications. In this sense the multidimensional situation contrasts sharply with the case of curves, where S. Stepanov, and later E. Bombieri, obtained an elementary proof of Weil's theorem.

Deligne's theory of "mixed Hodge structures," which develops cohomology theory for complex algebraic varieties with singularities, must also be mentioned. His papers significantly generalize classical results of W. Hodge, Kodaira, Serre, and others. The papers on Hodge theory, written almost simultaneously with the proof of the Weil conjecture, have profound internal connections.

The variety of papers Deligne has written in number theory and algebraic geometry is enormous. Let me point out just one recent result related to the study of hypergeometric functions. This area of mathematics, after being actively studied in the late nineteenth and early twentieth century, had lost its interest for pure mathematicians. The recent revival of the investigations in the theory of special functions arises from the papers of Deligne and G. Mostow and the new perspective of representation theory developed by the school of Gel'fand [DM] [GKZ].

Deligne and Mostow described the monodromy group Γ of a multi-dimensional hypergeometric function of the form

$$\int z^{\lambda}(z-1)^{\lambda_1}\prod_{i=1}^{d}(z-x_i)^{\lambda_i}\,dz.$$

For certain values of the parameters the group Γ forms a discrete lattice with finite covolume in the group of $d-1$-dimensional projective transformations, and in some cases nonarithmetic lattices are obtained. Together with the results of Gromov and I. Piatetski-Shapiro, who have constructed a whole series of nonarithmetic lattices in Lobachevskii spaces (spaces of rank one), these results complement the research of G. A. Margulis on arithmetic lattices in spaces of rank ≥ 2.

Gerd Faltings. The last work on post-Grothendieck algebraic geometry recognized by a Fields medal was a remarkable paper of Faltings proving the Mordell conjecture, which had been open for 60 years. A fundamental step was thereby taken on the road to proving Fermat's last theorem.

In simplified form the Mordell conjecture asserts that a system of algebraic equations with rational coefficients defining an algebraic curve of genus ≥ 2 has only a finite number of rational solutions. In proving the Mordell conjecture Faltings used results of S.Yu. Arakelov, Deligne, Manin, Mumford, A. Neron, A.N. Parshin, I.R. Shafarevich, J. Tate, and Yu.G. Zarkhin. This list is by no means complete.

A crucial difficulty overcome by Faltings in proving the Mordell conjecture arises from a fact common to many problems in algebraic number theory, and which seems remarkable at first glance. This difficulty is the Riemann problem for the zeta-function. The corresponding analogues of these theorems stated for function fields can be proved much more simply. In particular, as early as 1963 Manin obtained the analogue of Mordell's theorem for function fields.

In this regard Faltings outlined an entire program of research into algebraic varieties over number fields, which were called *arithmetic surfaces.* No wonder that Faltings' recent work in this area can be applied in physics in analyzing the multiloop contributions to a statistical sum of strings.

In selecting the medalists the committees have always tried to adhere to the basic principle of the founder of the prize—to reward both the solution of particular difficult long-standing problems and the formulation of new concepts that enlarge our knowledge. As far as the solution of specific classical problems is concerned, the medalists' papers in the area of analytic number theory can be assigned to this category in its purest form.

Number Theory

Atle Selberg. In 1950 Selberg became the first medalist honored for work in number theory. He developed an extraordinarily efficient method of estimating the distribution of primes. Since the time of Eratosthenes the sieve method has been applied to estimate the number of primes in a given interval. In 1919 Selberg's compatriot Norwegian mathematician V. Brun greatly improved the sieve method with his double sieve method, i.e., simultaneous estimation of the number of primes in two sequences. This method enabled him to obtain an estimate for the number $\pi_2(x)$, the number of twin prime pairs, each of which is less than x.

Brun's Theorem. *When* $x > x_0$

$$\pi_2(x) < \frac{cx(\ln\ln(x))^2}{\ln^2 x},$$

where c *and* x_0 *are positive constants.*

Selberg obtained significantly more precise estimates in the sieve method, enabling him to solve classical problems of number theory. The solution of one—finding an elementary proof of the asymptotic law for the distribution of primes

$$\pi(x) \sim x/\ln x \tag{8}$$

—resolved a paradoxical situation in number theory, which, in the words of G.H. Hardy, had thrown out a challenge to modern mathematics. Until Selberg's work the only way of proving the above formula (8), given by J. Hadamard and C.-J. de la Vallée-Poussin in 1897, relied on the theory of functions of a complex variable. Now thanks to Selberg and P. Erdös, who proposed another elementary version based on the Selberg formula, it was possible to obtain an entire proof within the framework of number-theoretic estimates. Selberg's results facilitated advances in solving a whole series of thorny problems in number theory.

Let me mention just one remarkable result of L.G. Shnirel'man.[16] Using original ideas on the density of the distribution of primes, he proved the following theorem in 1930:

Shnirel'man's Theorem. *Every integer can be represented as the sum of at most c primes, where c is an absolute positive constant.*

Shnirel'man obtained the estimate $c = 800,000$. By use of Selberg's technique, this number was reduced to 20 for sufficiently large numbers by H.N. Shapiro and R.S. Varga in 1951. The best known result is due to British mathematician R.C. Vaughan, who obtained the estimate $c_0 \leq 7$. An important detail needs to be pointed out at this

[16] The reasons for the suicide of the brilliant mathematician Shnirel'man (1905–1938) are not fully understood. Along with his outstanding papers in number theory, he is credited, together with L.A. Lyusternik, with solving Poincaré's problem on the number of distinct geodesics on an ellipsoid, the introduction of the concept of category of a manifold, and much more. If not for the forced isolation of Soviet mathematicians he might have been a sure candidate for the Fields prize. In the 1930s a group of world-class young mathematicians grew up in the USSR. It suffices to name A.N. Kolmogorov, Pontryagin, and Gel'fond. But the rupture in international scientific bonds, which was no fault of theirs, prevented their timely recognition.

juncture. Although Shnirel'man's method makes it possible to obtain the required representation for all numbers, a difference exists between the estimates for the quantity c_0 for all numbers and c_0' for sufficiently large numbers. Thus Vaughan obtains the result $c_0' \leq 7$ for sufficiently large numbers, while $c_0 \leq 27$ for all numbers. Even the result of Shnirel'man does not make it possible to solve the classical Goldbach problem—to obtain the desired estimate $c_0 \leq 3$. A known result of Vinogradov gives the estimate $c_0 \leq 4$ only for sufficiently large numbers. Later a lower bound for these numbers was obtained: $N \sim N_{6,60}$. A beautiful exposition of Selberg's method and the accompanying results appears in the book of Gel'fond and Yu.V. Linnik [GL].

Among the papers of Selberg that were so highly esteemed by the Fields Committee was his 1942 doctoral dissertation, which contains brilliant results on the problem of the distribution of zeros of the Riemann zeta-function. The Riemann conjecture is that all zeros of $\zeta(s)$ ($s = \sigma + it$) except for the trivial ones (-2,..., -2n,...) lie on the line $\operatorname{Re} s = \frac{1}{2}$. Despite efforts of many outstanding mathematicians, this conjecture remains unproved to the present. Of course any advance in this problem is regarded as a great achievement. Among the predecessors of Selberg are such famous mathematicians as Hadamard, H. von Mangoldt, E. Landau, Hardy, and Littlewood. In particular, in 1914 Hardy proved that there are infinitely many zeros on the line $\sigma = \frac{1}{2}$, and Littlewood constructed an estimate of the number of zeros $N_0(T)$ on the line $\sigma = \frac{1}{2}$ in the interval $0 \leq t \leq T$: $N_0(T) > \gamma T$. Mangoldt had previously proved Riemann's conjecture that the number of zeros $N(t)$ of the zeta-function in the critical strip $0 < \sigma < 1$, $0 \leq t \leq T$ is asymptotically given by $N(T) \sim \frac{T}{2\pi} \log T$. Selberg obtained an estimate of this type for the density of zeros on the critical line $\sigma = \frac{1}{2}$ itself. His result $N_0(T) > \gamma T \log T$ (where γ is a certain small constant) was not improved until American mathematician N. Levinson obtained the value $\gamma = \frac{1}{3}$ 30 years later.

Research in analytic number theory not only brought world wide fame to Selberg, it also was the source of remarkable new papers, which revealed unexpected links between number theory and other branches of mathematics. Selberg's ideas made it possible to combine areas of mathematics as seemingly remote from one another as the theory of discrete groups and automorphic forms, representations of semi-simple Lie groups, the theory of the zeta-function, scattering theory, and several others. Central to this body of problems was a 1956 paper in which Selberg obtained his trace formula. The problem that led Selberg to his formula is connected on the one hand with the study of real-analytic Eisenstein series, which are important in number-theoretic problems, and on the other hand with finding the spectrum of Laplace operators defined on symmetric spaces X of negative curvature. I shall sketch the group-theoretic construction that forms the basis of Selberg's result.

Suppose the semi-simple Lie group G is the group of motions of a space X, and Γ is a discrete subgroup of G such that the quotient space $\Gamma \setminus G$ is compact. Let $T(g)$ $(g \in G)$ be a unitary representation of the group G acting via shifts in the Hilbert space $L^2(\Gamma \backslash X)$ of functions that are square-integrable with respect to the invariant measure on $\Gamma \setminus X$. The representation $T(g)$ is reducible, and the basic problem of representation theory is to decompose it into irreducible components and find the multiplicities N_k of the irreducible representations. Since the operators $T(g)$ act on an infinite-dimensional space and have no trace in the usual sense, one considers instead of them the convolution operators of the form

$$T\varphi = \int \varphi(g)T(g)\,dg.$$

Under certain natural conditions on the function φ, for example, assuming that φ are functions of compact support, the operator $T\varphi$ is a nuclear operator for which the concept of a trace is defined in the ordinary sense as the sum of the eigenvalues of the matrix kernel of the operator $T\varphi$. The following trace identity holds:

$$\int\limits_{\Gamma\backslash X} \sum_{\gamma\in\Gamma} \varphi(g^{-1}\gamma g)\,dg = \sum_{k=1}^{\infty} N_k \int\limits_{G} \varphi(g)\sigma_k(g)\,dt. \qquad (9)$$

Here $\sigma_k(g)$ are the characters of the irreducible representations occurring in $T(g)$, and N_k are the multiplicities with which they occur in that representation.

The left-hand side of the identity (9) can be brought into the form

$$\sum \mu(\Gamma_\gamma \setminus G_\gamma) \int\limits_{G/G_\gamma} \varphi(g^{-1}\gamma g),$$

where the summation extends over the set of conjugacy classes of the group Γ. Here Γ_γ and G_γ denote the centralizers of the element γ in Γ and G respectively, and

$\mu(\Gamma_\gamma \setminus G_\gamma)$ is the measure of the space $\Gamma_\gamma \setminus G_\gamma$. The principal problem is to obtain explicit formulas for the integral $\displaystyle\int \varphi(g^{-1}\gamma g)\, dg$.

Selberg found explicit formulas for the traces of the operators $T\varphi$ for several important classes of symmetric spaces (spaces of rank one). These formulas are expressed on the one hand as sums of a series over the conjugacy classes of the elements of the group Γ, and on the other hand as functions of the eigenvalues of the invariant Laplace operators defined on X and acting in the space of representations of the group G. The Selberg trace formulas have most important applications in computing the dimensions of the space of automorphic forms, analysis of Eisenstein series, the study of multi-dimensional zeta-functions, and the investigation of other problems of the theory of representations and number theory. Indeed, number-theoretic research led Selberg to his famous result. In particular, the classical Riemann hypothesis on the zeros of the zeta-function admits an interpretation in terms of the trace formula. Another predecessor of the Selberg formula is the classical Poisson summation formula:

$$\sum_{n\in\mathbb{Z}} e^{in\varphi} = \sum_{m\in\mathbb{Z}} \delta(\varphi - 2\pi m). \tag{10}$$

An elementary, though very useful exercise for the reader would be to analyze the Poisson summation formula from the group-theoretic point of view, taking into account that the role of G is played by the real line $\mathbb{R}$, while Γ is the group of integers $\mathbb{Z}$. Formula (10) and its generalizations are important in physics, especially in the theory of phase transitions. Formulas of type (10) make it possible to find the points of phase transition in several models of statistical physics, for example in the Ising model.

Selberg originally obtained his formula by considering the groups $G = \mathrm{SL}\,(2, \mathbb{R})$ and $\Gamma = \mathrm{SL}\,(2, \mathbb{Z})$. The group G is the group of rigid motions of the hyperbolic plane (the upper half-plane) presented in the form $\mathrm{SL}\,(2, \mathbb{R})/\mathrm{SO}\,(2)$. The irreducible representations of the group $\mathrm{SL}\,(2, \mathbb{R})$ can be realized in the space of eigenfunctions of the Laplacian on the hyperbolic plane. The realization of the Selberg formula for the group $\mathrm{SL}\,(2, \mathbb{R})$ and its discrete subgroups leads to some very interesting questions. Please note that the space $\mathrm{SL}\,(2, \mathbb{Z})/\mathrm{SL}\,(2, \mathbb{R})$ is noncompact, but has finite volume. Obtaining the trace formula in this case is far more difficult than in the compact case, since the spectrum of the operators of the representation has a continuous part. Another body of questions is related to the description of discrete subgroups Γ having the property $\mu(\Gamma \setminus G) < \infty$. Among the groups having this property there is the intriguing subclass of arithmetic groups, which Weil described completely for subgroups of $\mathrm{SL}\,(2, \mathbb{R})$. Examples are the group $\Gamma = \mathrm{SL}\,(2, \mathbb{Z})$, the quaternion groups, and their subgroups of finite index. The arithmetic property of discrete groups is very important in the multi-dimensional situation. Not having time to discuss even a small part of the applications of the Selberg trace formula, I refer the reader to corresponding literature [He1, He2, Ve].

In studying properties of discrete subgroups of Lie groups Selberg stated many important conjectures on their structure. The following generations of mathematicians proved some of them. Another Fields medalist, Margulis, obtained the chief results in this direction.

Klaus Roth. In 1958 Roth was honored for the proof of a delicate estimate that refines the Thue-Siegel theorem on the approximation of algebraic numbers by rational numbers. He proved the following theorem.

Roth's Theorem. *If α is any algebraic number, not itself rational, then for any $\nu > 2$ the inequality*

$$\left|\frac{p}{q} - \alpha\right| < \frac{1}{q^\nu}$$

has only a finite number of solutions in rational p/q.

This theorem is best-possible. It would not remain valid if $\nu = 2$. Among the important developments to which Roths' work gave rise was a far-reaching generalization on simultaneous approximations obtained by W.M. Schmidt in 1970. Termed the subspace theorem, it furnishes best-possible multi-dimensional results and completely answers the question of which normal form equations have only finitely many solutions. Faltings and G. Wustholz have recently profoundly extended this work in the context of algebraic geometry. Incidentally, in another direction, P. Vojta has utilized algebraic geometric ideas and some classical work in the Thue-Siegel context from Dyson to develop a new proof of Faltings' celebrated result on the Mordell conjecture. Faltings has extended Vojta's method to construct a theory of Diophantine approximation on algebraic (more precisely, abelian) varieties.

Another series of papers by Roth deals with the problem of P. Turán and Erdös, already known in the 1930s, on sequences containing arithmetic progressions.

Let $r_k(n)$ be the smallest integer ρ for which each sequence of positive integers $a_1 < a_2 < \cdots a_\rho \leq n$ contains an arithmetic progression with k terms. More than 60 years ago Turán and Erdös conjectured that $r_k(n) = o(n)$.

Roth obtained the first substantial result, the proof of the Turán-Erdös conjecture for $k = 3$. Quite recently Hungarian mathematician E. Szemerédi proved the conjecture completely [Sze]. Szemerédi's work attracted the attention of specialists in ergodic theory. H. Furstenberg, Y. Katznelson, and D. Orenstein then found a new proof of Szemerédi's theorem and obtained a multi-dimensional generalization of it [FKO].

Apparently the latter proof does not give sharp estimates for $r_k(n)$. The result of Roth and F. Behrend for $r_3(n)$ long remained optimal:

$$\frac{n}{e^{c_1\sqrt{\log n}}} < r_3(n) < \frac{c_2 n}{\log\log n}.$$

Szemerédi later obtained the better upper bound $r_3(n) < \dfrac{c_2 n}{(\log n)^\gamma}$ for some small $\gamma > 0$.

In 1966 Roth published a monograph in collaboration with H. Halberstam which is still a very good introduction to elementary methods in analytic number theory, especially combinatorial problems and sieve methods [HR].

In 1965 Roth discovered an optimal result in conjunction with the large sieve of Linnik and A. Renyi. Bombieri and A.I Vinogradov subsequently took up this field of research with fruitful consequences.

Alan Baker. The work of Baker, who was a 1970 prize winner, continues that of the brilliant members of the Cambridge school of number theory—Hardy, Littlewood, and

Ramanujan. Baker developed a powerful method for estimating linear forms in the logarithms of algebraic numbers. The impact of his work on many classical questions is far-reaching. An early application was the solution to the problem of K. Gauss, whether there are only nine imaginary quadratic fields with class number one. Baker established the result through an argument going back to Gel'fond and Linnik. Remarkably, H. Stark simultaneously gave another verification, motivated by a paper of Heegner. The solution of J. Ax and A. Brumer to the Leopoldt problem on the p-adic regulator of an abelian field was another early success directly dependent on Baker's work. But by far the most profound application has been to the theory of Diophantine equations. Baker first effectively demonstrated Thue's theorem on the representation of integers by binary forms. By a different method, he had previously dealt with special Thue equations of the form

$$x^3 - ay^3 = n,$$

and these results greatly improved the classical work of B.N. Delone (1922) and T. Nagell (1925), who had found a complete solution only in the case $n = 1$. Returning to logarithmic forms, Baker showed that they underpin both the theoretical and the practical determination of all the integer points on a wide range of Diophantine curves. These curves include the one defined by the Mordell equation

$$y^2 = x^3 + k$$

and indeed any equation of hyperelliptic or superelliptic type. Subsequent refinements in the fundamental estimates by Baker and others have spawned the effective solution of a whole new class of examples, termed exponential Diophantine equations. In these instances not even an ineffective theory existed previously; that is, the number of solutions was not even known to be finite. As R. Tijdeman demonstrated, the new class includes the Catalan equation

$$x^p - y^q = 1,$$

which is the subject of a recent book by P. Ribenboim [Ri].

Baker's methods originated in the classical studies of Gel'fond, C. Siegel, and T. Schneider that led to solving Hilbert's seventh problem on the transcendence of numbers of the form α^β, with α algebraic and β algebraic irrational. The key result here is the following theorem.

Baker's Theorem. *If* $\alpha_1, \ldots, \alpha_n$ *are nonzero algebraic numbers such that* $\log \alpha_2, \ldots, \log \alpha_n$ *are linearly independent over the rationals, then* $1, \log \alpha_1, \ldots, \log \alpha_n$ *are linearly independent over the field of all algebraic numbers.*

This theorem has been generalized, principally in the context of algebraic groups. Profound connections have been discovered with complex function theory, Kummer theory, and many aspects of Diophantine geometry, including Faltings' theorem on the Mordell conjecture, to which we referred earlier. Without doubt it is an active area of research, and a book covering the subject is in preparation by Baker and Wustholz. A good idea of some earlier material is provided in the treatise of T.N. Shorey and Tijdeman [Sh] and a collection of articles [Ba].

Among Baker's outstanding achievements is the 1966 solution of a problem that goes back to Gauss: Find all imaginary quadratic fields $Q(\sqrt{d})$ (i.e., sets of numbers of the form $u + v\sqrt{d}$, where u and v are rational numbers) having the property of unique decomposition, up to order, into prime factors. Baker showed that there are nine such fields: $d = -1, -2, -3, -7, -11, -19, -43, -67, -163$. His proof is based on his theory of transcendental numbers. H. Stark reached this result simultaneously and independently of Baker by a different method, using elliptic functions. In the case of real fields ($d > 0$), however, the solution of this problem remains unknown. It is unknown, for example, whether infinitely many such fields exist.

Enrico Bombieri. The last work to be noted in this list of number theorists is a paper of Bombieri, a 1974 prize winner. Bombieri especially investigated the large sieve, finding an inequality which shows that on the average the Riemann hypothesis holds with respect to primes in arithmetical progressions. About the same time, A.I. Vinogradov independently obtained a quite similar result. An immediate application of this work arising from the classical work of Renyi shows that every large even integer is the sum of a prime and a number with at most three prime factors. This result was the best available on the Goldbach conjecture to that time. Later, J. Chen ingeniously added the idea of replacing "three factors" by "two factors." For further discussion see [Da]. Bombieri also authored a large series of papers on the implementation of the Thue-Siegel-Roth method.

Bombieri's work extends beyond number theory. An exceptionally versatile scholar, he is the author of many first-class papers in algebraic number theory, classical analysis, algebraic geometry, and quasi-crystals. I shall first discuss his result, obtained in 1969 jointly with E. di Giorgi and E. Giusti, the multi-dimensional analogue of the theorem of S.N. Bernstein from the theory of minimal surfaces.

Bernstein's theorem has a long history. In 1902 Bernstein proved that a complete two-dimensional regular surface with zero mean curvature in $\mathbb{R}^3$ is a plane. The conjecture arose that this result holds in any dimension. In 1968 J. Simons proved Bernstein's conjecture for dimension ≤ 7. For $M^8 \subset \mathbb{R}^9$, however, the analogue of this theorem is false.

In the same paper Simons identified another class of locally minimal surfaces in $\mathbb{R}^{2m}$—the locally minimal cones. Simons' cones are defined by a system of equations in $\mathbb{R}^{2m}$:

$$x_1^2 + \cdots + x_m^2 = x_{m+1}^2 + \cdots + x_{2m}^2 < r^2.$$

They are located in the sphere S^{2m} of radius r and have boundary $S_r^m \times S_r^m \subset S^{2m}_{\sqrt{2}r}$; (here S_r^m denotes a sphere of radius r in $\mathbb{R}^m$).

As shown in a 1969 paper of Bombieri, di Giorgi, and Giusti, Simons' cones give a global minimum for $n \geq 8$. The subtle proof makes clever use of the Bendixson-Poincaré theory at the final stage.

Bombieri understood immediately the remarkable prospects opened up in solid state physics by the discovery of quasi-crystals in 1984 (see [SBGC]). Quasi-crystals are substances that have local symmetries of order five (icosahedral groups) and, in contrast to ordinary crystals, do not admit translation invariance. The discovery of quasi-crystals shed new light on papers of R. Penrose, N.G. De Bruin, R.M. Robinson, and other mathematicians on quasi-crystal tilings of a plane, and they provoked a flood of papers. It was shown that the most natural method of constructing quasicrystal tilings of the plane and three-dimensional space $\mathbb{R}^3$ is the "cut and project" method, i.e., projection of points of a regular n-dimensional lattice ($n = 5$ for the plane and $n = 6$ for $\mathbb{R}^3$) onto an "irrationally" imbedded subspace $\mathbb{R}^2$ or $\mathbb{R}^3$ respectively (imbedded so that the only point with integer coordinates it contains is the origin). In his paper "Quasi-crystals, tilings, and algebraic number theory," written jointly with J. Taylor [BT], Bombieri linked the problem of classifying quasi-crystals with problems of number theory. They studied the problem of constructing quasi-crystal tilings starting from special local rules for gluing together elementary cells. The major result of this paper was a construction using special properties of algebraic numbers (the elements of Galois theory and Pisot numbers) and quasicrystals, which cannot be obtained by the "cut and project" method. At approximately the same time S.P. Novikov, who had become interested in the problem of classifying quasi-crystals, was proposing the concept of quasi-crystallographic group, a nontrivial generalization of the concept of the usual crystallographic group. His student S. Piunichin obtained a fairly complete classification of the two-dimensional quasi-crystallographic groups and constructed new examples of quasi-crystal tilings not possible by the intersection and projection method. To solve this problem, Piunichin applied the elements of algebraic K-theory. Undoubtedly the theory of quasi-crystals will pose new and unexpected problems for mathematicians.

Jesse Douglas. The mathematical community has always seen the solution of a classical problem in the theory of minimal surfaces as an outstanding achievement in general mathematics. The winner of the first Fields prize, Douglas, was honored for his solution of Plateau's problem. Largely through his 1847 soap film experiments, the Belgian physicist, J. Plateau created a new area of research—the theory of minimal surfaces. But nearly 90 years passed before a rigorous mathematical proof appeared in works of Douglas and T. Radó. Radó's proof depended on the conformality properties of two-dimensional surfaces. Douglas' more general proof contained ideas that were essential in later investigations. The statement of Plateau's problem is well known: *Prove that for a given Jordan curve* $\Gamma \subset \mathbb{R}^n$ *there exists a surface of minimal area having* Γ *as its boundary.* This problem, which is intuitively clear when the boundary Γ is simply a circle, becomes nonobvious when one considers a complicated curve Γ, for example, a system of linked circles. W. Fleming (cf., for example, [L1]) showed that it is possible to construct a system of minimal surfaces of various topological types, having linked circles as their boundary. The surface having an absolute minimum area is unique among all surfaces of fixed topological type (e.g., simply connected surfaces) however, as the solution of Plateau's problem implies.

Sixty years after its publication Douglas' paper gained attention through the development of the theory of strings. Italian physicist T. Regge pointed to the connection between the Douglas functional that is being minimized and the ground state of the wave function of a string. Following Regge and omitting technical details, I shall exhibit this analogy. Consider the simplest case of Plateau's problem, in which the surface S is diffeomorphic to the unit disk $\mathcal{D} \subset \mathbb{R}^n$ and bounded by the curve Γ. Consider all the parametrizations of Γ, i.e., all mappings $g : \mathbb{S}^2 \to \mathbb{R}^n$, $x = g(\sigma)$, where $0 \le \sigma \le 2\pi$ and $g^\mu(0) = g^\mu(2\pi)$.

Douglas mainly studied the integral

$$A(g) = \frac{1}{4\pi} \int\limits_{\mathbb{S}^1} \int\limits_{\mathbb{S}^1} \frac{|g(\sigma) - g(\theta)|}{(2 \sin \frac{1}{2}(\sigma - \theta))^2} \, d\sigma \, d\theta.$$

The integrand has a simple geometric meaning: the ratio of the square of the length of chords between points on the curve Γ and their preimages in $\mathbb{S}^1$. The functional $A(g)$ is a minimum for some mapping g^*. If the function g^* is continued to the interior of $\mathfrak{D}$ by the classical Poisson formula, we obtain $x = \operatorname{Re} F(w)$, where $w = u + iv$. When this procedure is done,

$$\sum_{i=1}^{n} F_i'^2(w) = 0. \tag{11}$$

As early as 1865 K. Weierstrass showed that the condition (11) defines a minimal surface S^*. The area of S^* is $A(g^*)$. To establish the connection with quantum oscillations of a string, it suffices to expand $g(\sigma)$ as a Fourier integral

$$a_m = \frac{1}{2\pi} \int\limits_{\mathbb{S}^1} g(\theta) e^{-im\theta} \, d\theta.$$

In the context of this expansion the functional $A(g)$ can be expressed as

$$A(g) = 2\pi \sum_{m=1}^{\infty} m |a_m|^2.$$

The wave function $\Phi = \exp(-A(g))$ includes all the natural vibrations of the string. The physical consequences of this analysis are given in the article of Regge [Re].

An important part of Douglas' theorem is the assertion that the parametrization g^* that minimizes $A(g)$ also defines the minimal surface. V. Guillemin, B. Kostant, and S. Sternberg gave an elegant proof of this assertion using the invariance of the functional $A(g)$ with respect to all reparametrizations of the mapping g from the group $SL(2,\mathbb{R})$ [GKS].

N. Hitchin offered another interesting observation connecting the theory of minimal surfaces with modern physical structures. He remarked that the Weierstrass criterion, which is based on the introduction of a complex structure on the minimal surface, admits a natural interpretation in the language of Penrose twistors.

Algebra

John Thompson. The classification of the finite simple groups is a classical problem of mathematics. At present this topic seems completely solved. Thompson, a 1970 prize winner, made an important contribution to this classification.

Moving ahead, I should explain why the phrase "seems to be" occurs in a precise mathematical assertion. D. Gorenstein, a leading expert in the theory of finite groups, used this phrase. It was evoked by a rather unusual problem in the history of mathematics.

The proof of the completeness of the classification contains about 5000 journal pages. Moreover a thorough understanding requires an equal number of supplementary pages, since some results were obtained using a computer. The verification of the proof is a difficult problem itself. The book of Gorenstein [Gor] gives a partial exposition, and two other volumes by Gorenstein have appeared with detailed proofs.

Classifying the simple finite groups is immeasurably more complicated than, for example, classifying the simple Lie algebras. Among the finite groups there are 26 exceptional groups, and the orders of these groups can be quite large. For example, the order of the maximal sporadic group, the Fischer-Griess group, known according to taste as either the *monster* or the *friendly giant*, is

$$2^{46}\cdot 3^{20}\cdot 5^{9}\cdot 7^{6}\cdot 11^{2}\cdot 13^{2}\cdot 17\cdot 19\cdot 23\cdot 29\cdot 31\cdot 41\cdot 47\cdot 59\cdot 71 \sim 10^{54}.$$

This group can be represented as the group of automorphisms of a certain nonassociative but commutative algebra of dimension 196,883. Some wonderful recent discoveries involve this group, which is discussed below.

Any finite simple group is either of Lie type, i.e., the analogue of a Lie group over a field of finite characteristic, or an alternating group A_n ($n \geq 5$), or one of the exceptional (sporadic) groups. The proof that the sporadic groups are simple requires a special technique. The compilation of a final list of sporadic simple groups completed the classification, but the preceding stage was also difficult: the proof of a number of theorems on the structure of simple groups. These theorems provide a way to find common regularities in the structure of finite simple groups. In this direction the results of Thompson are most important. In a joint paper with W. Feit he proved the fundamental result: all nonabelian simple groups are of even order.

Thompson's papers cover the entire subject of finite groups. He has continued to work actively in this area. One of the problems he posed was recently solved by Bombieri.

The main developments in the theory of finite simple groups are connected with the study of the Fischer-Griess monster. Thompson and J. McKay discovered that the dimensions of the representations of the monster are the coefficients in the expansion of the modular function $J(\tau)$ defined by the Dedekind eta-function and the theta-function of the weighted lattice of the simple Lie algebra E_8. This observation directly linked the study of the Fischer-Griess group and work on infinite-dimensional Lie algebras, which developed in parallel with it. The work of many mathematicians, especially I. Macdonald and V. Kac, went into the discovery of relations between the dimensions of representations of infinite-dimensional Lie algebras and identities for eta-functions [Kac]. A new area of mathematics had arisen, with completely unexpected results and applications, including such seemingly widely separated areas as string theory, two-dimensional conformal theories in physics, the classification of Leech lattices, coding theory, and more. Two recently published beautiful books [FLM, CS] give a clear presentation of these results.

Reviewers of the papers of Fields medalists constantly encounter the difficulty of assigning a given paper to some traditional area of mathematics, so much has the face of mathematics changed over the last 30 years. To which branch of mathematics, for example, should the papers of Margulis and D. Quillen be assigned?

Miscellany

Grigorii Aleksandrovich Margulis. The most important of Margulis' results of is his proof of Selberg's conjecture that a certain class of discrete groups is arithmetic. While the conjecture can be stated rather easily, its proof required a virtuoso mastery of the technique of the theory of algebraic groups, use of the multiplicative ergodic theorem, the theory of quasi-conformal mappings, and much more. French mathematician J. Tits, in presenting one of Margulis' papers at the Helsinki congress, said: "During the year in which I conducted a seminar on the papers of Margulis, I learned more mathematics than in all the years preceding."

The theory of discrete groups is tied closely to the theory of Riemann surfaces. F. Klein and Poincaré had discovered that the description of Riemann surfaces of genus larger than one can be reduced to the study of the discrete subgroups of the group $SL(2,\mathbb{R})$ (the study of the unimodular group $SL(2,\mathbb{Z})$) acting on the upper half-plane $\operatorname{Im} z > 0$ as fractional-linear transformations $z \mapsto \frac{az+b}{cz+d}$). The discrete subgroup $\Gamma = SL(2,\mathbb{Z})$ is the subgroup of matrices in $SL(2,\mathbb{R})$ with integer coefficients. Other discrete subgroups in $SL(2,\mathbb{R})$ are subgroups of finite index in $SL(2,\mathbb{Z})$. As is well known, all Riemann surfaces of genus $g > 1$, both compact and noncompact, can be obtained by taking the quotient of $SL(2,\mathbb{R})$ over Γ_n, where Γ_n is a discrete subgroup in $SL(2,\mathbb{Z})$.

The two-dimensional theory can be generalized to the multi-dimensional case in various directions. One natural problem can be stated as follows: *Describe all the discrete subgroups of the group $SL(n,\mathbb{R})$ with the property that the* *space* $SL(n, \mathbb{R})/\Gamma$ *has finite volume (with respect to a given invariant measure).* C. Hermite proved that the space $SL(n, \mathbb{R})/SL(n, \mathbb{Z})$ has finite volume. Many subtle facts from algebraic number theory can be reduced to general statements of this type. Without giving precise definitions, I remark that the discrete subgroup Γ_n is an important example of the so-called arithmetic subgroups.

Arithmetic subgroups of semi-simple Lie groups G have the property $\mu(G/\Gamma) < \infty$, where μ is the volume of G/Γ with respect to an invariant measure. Margulis proved that under certain conditions on the rank of G ($\operatorname{rk} G \geq 2$) the converse is also true: every discrete subgroup Γ such that $\mu(G/\Gamma) < \infty$ is arithmetic. Various generalizations of this result and a number of beautiful corollaries have been obtained by many mathematicians over the last few years in the theory of discrete groups and related areas of mathematics. This work attests to the profundity of Margulis' theorems.

Margulis has also arrived at some beautiful results in such areas as ergodic theory and the theory of foliations. Margulis' papers display his exceptional originality. By invoking ideas from widely separated branches of mathematics, he proceeds to the goal by the simplest possible route.

One of these theorems, which was central to proving the theorem on arithmetic subgroups, was announced several years before the full proof appeared. Despite efforts of many leading experts in this area, the result could not be reproduced independently. The nature of Margulis' original proof was very simple and, to the experts, unexpected [Marg1].

In recent years Margulis has examined the properties of discrete groups in different and sometimes unexpected areas. For example, he refuted a conjecture of Milnor on the structure of groups of motions acting discontinuously on affine spaces. It had been believed that such groups are always solvable (polycyclic). Margulis constructed a counterexample. By combining ideas from the theory of discrete groups and ergodic theory, he recently solved an old problem of the geometry of numbers—Oppenheim's problem on the representation of numbers by indefinite quadratic forms [Marg2].

Daniel Quillen. The 1978 Fields prize winner Quillen is a leading expert in algebra and algebraic topology. His achievements are extremely difficult to present in a brief form accessible to the nonspecialist. One of Quillen's impressive accomplishments was the proof of Serre's conjecture on the structure of projective modules over a ring of algebraic functions. At the same time, Leningrad mathematician A.A. Suslin obtained this result independently. The rapidly developing Soviet—now Russian—reality is difficult to convey. When this book was written, Suslin was a Soviet mathematician living in Leningrad. The Soviet Union has now broken up, Leningrad has become St. Petersburg, and Suslin is working at Northwestern University in the USA, nevertheless retaining his position in the Steklov Mathematical Institute.

J.-P. Serre stated this conjecture in 1955 in a now classic paper [S1]. This paper first extensively applied homological algebra to the study of algebraic varieties.

Serre's Conjecture. *Every projective module over a ring of polynomials is free.*

This assertion can be understood intuitively by an analogy with vector bundles. I shall illustrate it by an example proposed by Atiyah [At]. Consider the Möbius band M, representing it as a twisted (infinitely wide) cylinder. M can be thought of as a family of lines M_θ parameterized by the parameter θ of the base circle $\mathbb{S}^1$. Each such line forms a one-dimensional vector space, but there is no way of choosing a natural basis in it. On the other hand, the normals to M form an analogous linear bundle $M^\perp$ over $\mathbb{S}^1$. The direct sum $M + M^\perp$ is a two-dimensional vector bundle over $\mathbb{S}^1$. It can be regarded as the normal bundle to the middle line l of the Möbius band imbedded in $\mathbb{R}^3$. Since l is an ordinary plane circle, its normal bundle is trivial, and consequently in it a global basis can be introduced. We have thus represented a nontrivial one-dimensional bundle M over $\mathbb{S}^1$ as a direct summand of a trivial but two-dimensional bundle. The essence of Serre's conjecture is that by a suitable choice of the polynomial ring—the analogue of the coefficient field—a projective module (a nontrivial vector bundle) can be made free, i.e., a global basis can be introduced into it.

In an equally well-known development, Quillen proved J. Adams' conjecture in topological K-theory. This result was obtained at the same time by Sullivan, using a different method.

All of Quillen's results fall in the context of the construction of algebraic K-theory. His transference of Grothendieck's K-theory into a purely algebraic situation leads to the solution of many fundamental problems of algebra and number theory. Quillen even solved one in algebraic K-theory, the construction of the higher analogues of Grothendieck rings.

Quillen's recent paper [Q] gives an expression for the metric of a linear holomorphic bundle over the moduli spaces of Riemann surfaces. This result is important in modern investigations in string theory. In this theory integration over the moduli space emerges in computing statistical sums, correlation functions, and other basic objects of the theory.

Laurent Schwartz. For those who apply the subject, mathematics is primarily thought of as analysis. But relatively few analysts appear in the list of medalists. Besides Hörmander, a specialist in differential equations, medalists include Schwartz and C. Fefferman.

Schwartz, a 1950 prize winner, received the prize for his work in the theory of generalized functions (distributions). The theory of generalized functions originated in the work of Hadamard, M. Riesz, N.M. Gunther, S.L. Sobolev, and Dirac.[17] Its finished form and wide applications may be traced to Schwartz.

Schwartz regarded a distribution as a functional on a space of test functions. In this sense his approach closely resembled the definition of Dirac. Such a technique made it possible to expound all basic problems of the theory of distributions from a unified point of view—the theory of

---

[17] As recently discovered, the Dirac delta-function had appeared in the works of O. Heaviside at the end of the nineteenth century. Some idea of the reaction of pure mathematicians to Heaviside's work can be gained from Hardy's, *Divergent Series*.

integration, differentiation, Fourier transform, and so on. In the apt phrase of Dieudonné, one can say that, just as Leibniz and I. Newton did not discover the differential and the integral, but proposed a system of calculus, so Schwartz constructed a calculus of generalized functions. His two-volume *Théorie des distributions* (1950), together with the six-volume course *Generalized Functions*, written by Gel'fand and co-authors, became the bible of specialists in functional analysis.

Schwartz' other remarkable achievements in the theory of nuclear spaces and complex manifolds were eclipsed by his papers on the theory of distributions. Such, alas, is the lot of those who have conquered a real summit.

Lars Hörmander. A 1962 Fields prize was awarded to Hörmander. Among the important topics Hörmander studied up to the time he received the prize was the general theory of linear partial differential equations constructed by him. Study of the relations between the smoothness of the coefficients of a differential operator P and the smoothness of the solutions is a significant theme in the theory of differential equations. One theorem of this type is the Cauchy-Kovalevskaya theorem, which is included in all textbooks. A more complicated problem was that of describing the classes of differential operators for which the problem has only analytic solutions:

$$Pu = 0 \Rightarrow u \text{ is an analytic function.}$$

Hilbert included this problem in his list of problems posed at the 1900 International Congress of Mathematicians in Paris. In the 1940s I.G. Petrovskiĭ solved this problem in the exact form stated by Hilbert. Petrovskiĭ proved many other general results in the theory of partial differential equations. In particular, he showed that solutions of nonlinear elliptic systems with analytic coefficients are analytic and stated the problem of constructing a general theory of systems of linear operators. This theory was the one that Hörmander constructed.

Hörmander extended Petrovskiĭ's theory by describing a class of differential operators P such that

$$Pu = 0 \Rightarrow u \in C^{\infty}. \tag{12}$$

The identification of a new important class of operators called *hypoelliptic* operators followed. The solution of problem (12) answered a question posed by Schwartz in his *Théorie des distributions.* Later work by Hörmander was associated with the theory of pseudodifferential operators, where he also derived fundamental results. His four-volume monograph on differential operators is an encyclopedic exposition of the subject [Hö].

Charles Fefferman. Fefferman, a 1978 prize winner, is known for reviving the study of the classical problems of analysis. He obtained strong results in real and complex analysis, solving the duality problem for the Hardy spaces and studying the convergence of Fourier series of functions of several variables. In these problems he also solved the multiplier problem for the ball [Fe]. Define an operator T in $L^p(\mathbb{R}^n)$ by the relation $\widehat{Tf}(x) = \chi_B(x)\hat{f}(x)$, where χ_B is the characteristic function of the unit ball, and the hat denotes the Fourier transform. Is the operator T bounded in the norm of $L^p(\mathbb{R}^n)$? Earlier it had been proved that the operator T is bounded if the ball is replaced by a cube. Fefferman's result for the ball is thus all the more surprising: *The operator T is bounded only in $L^2(\mathbb{R}^n)$ if $n > 1$.* Fefferman's proof ingeniously uses a beautiful classical construction of A. Besicovitch. I recall Besicovitch's result here for the reader. In 1917 the Japanese mathematician S. Kakeya stated the following problem: *In the class of figures in which a segment of length one can be turned through a complete revolution, while always remaining within the figure, which one has the smallest area?* Besicovitch's astonishing solution, obtained in 1928, was that the area of such a figure can be arbitrarily small. Subsequently the construction of analogues of such sets for regions in $\mathbb{R}^n$ $(n > 2)$ were critical in a number of problems of the theory of functions, in particular, in Bourgain's work on estimating oscillatory integrals.

Fefferman's papers on the classification of biholomorphic domains is also very interesting. This work builds upon sharp estimates of the Bergman kernels. Fefferman offers a harmonious blend of the methods of real and complex analysis. Especially beautiful are his results on real hypersurfaces and complex manifolds.

Alain Connes. The work of Connes, a 1983 prize winner, fits into functional analysis with a bit of stretching. His principal results are in the theory of factors created by J. von Neumann in the theory of operators. In the 1930s and 1940s von Neumann and F. Murray laid the foundations of factor theory.

The basic propositions of this theory, whose source is in the commutation relations of quantum mechanics, reduce to the following. Let A be a ring of operators acting in a Hilbert space H, and A^* the ring of operators that commute with A. The rings A and A^* form a *factor* if $A \cap A^* = \{\lambda E\}$, where E is the identity operator. In the finite-dimensional case the rings A are isomorphic to rings of matrices acting in a space $\mathbb{R}^n$. This result follows from Schur's lemma. The dimension n is the only invariant of a factor A.

The infinite-dimensional case involves an immeasurably more complicated situation. Murray and von Neumann introduced the concept of relative dimension (Δ), which makes it possible to classify factors in the infinite-dimensional case. Besides factors of the type described (which are said to be of type I) there are two other classes of factors called types II and III. Factors of type II are divided into two subclasses II_1 and II_∞. In class II_1 the quantity Δ can assume any value in a finite interval $[0, \lambda)$ and in II_∞ any value in $[0, \infty]$. Here λ is a positive real number. For class III the quantity Δ assumes only the two values 0 and ∞. In the 30 years following the papers of von Neumann and Murray almost no advances were made in factor theory. Then, in the 1960s, the situation changed. Many mathematicians contributed to advancing the classification of factors of types II and III. In particular, in 1967 R. Powers constructed a continuous family of pairwise non-isomorphic factors of type III. Connes completely classified factors of type III and solved a series of problems in factor theory posed in the foundational papers of von Neumann. He also found new, unexpected applications of this theory.

Connes has also conducted a promising cycle of research in a new area—noncommutative differential geometry.

Concerning the papers of Connes, I wish to make some general remarks. The fate of the papers of von Neumann and Murray in factor theory is rather curious. They were inspired by von Neumann's deep interest in the problems of quantum mechanics, as one can tell from the text of articles. It was assumed that factors would be applied in quantum field theory and statistical physics. But, except for some highly artificial, desultory applications, this application has not yet happened. Recent advances in field theory, especially two-dimensional conformal theories, may change the situation.

This hope has been reinforced by some remarkable recent discoveries. V. Jones contributed the main result when he constructed a new class of invariants of knots of polynomial type. The Jones invariants differ from those of Alexander, which have been known since the 1920s and provided a basis for solving some classical problems in knot theory. Perhaps no less remarkable than the result itself is the method of constructing such invariants. Jones polynomials are closely connected with braid groups and Hecke algebras that generate factors of type II_1. In addition, such algebras arise in many models of statistical physics. Immediately after the papers of Jones, whole series of new polynomial invariants were obtained, defined by exactly solvable statistical models. In this field of research, mathematicians are discovering huge numbers of unexpected connections among topological, field-theoretic, and group problems. E. Witten [Wi2] has recently developed a fascinating approach to the construction of knot theory by field-theoretic methods.

Although I am compelled to leave this theme, let me point out an amazing fact that becomes evident when the classification of conformal theories and factors in Jones' theory are compared. One nearly identical number characterizes both—the central charge in the first case and the factor index in the second [BPZ, J1].

Connes' work on noncommutative differential geometry has recently been applied to and connected with superstring models. Witten has applied these subtle results to derive the Lagrangians for superstrings. In another recent cycle of research applying noncommutative algebraic geometry to solid state physics, J. Bellisard attempts to explain the quantum Hall effect ([Be]). Connes has also investigated applications to physics. He has developed with enthusiasm the concept of physical space-time represented not as a set of points, but as a noncommutative space [Con].

These unforeseen ways of applying very abstract mathematical structures illustrate the "unreasonable effectiveness of mathematics in the natural sciences."

Mathematical Logic

Paul Cohen. The contributions of Cohen to mathematical logic—the solution of the continuum hypothesis—are unusual. G. Cantor stated the continuum hypothesis in 1878. A later formulation of it runs as follows: *There does not exist a set of cardinality intermediate between a countable set and a set having cardinality of the continuum.* In concluding an 1884 paper in the *Mathematische Annalen* with the words "to be continued," Cantor planned to prove the continuum hypothesis in the next part of the paper. The planned paper never appeared, however, and thereby shared the fate of many famous papers, beginning with Fermat's proof of his last theorem.[18]

Despite many attempts, several generations of mathematicians failed to prove or refute the continuum hypothesis. From 1938 to 1940 K. Gödel obtained an outstanding result in relation to this question, when he proved that the continuum hypothesis is consistent with the axioms of set theory, including the axiom of choice. Finally, in 1963 Cohen proved that the continuum hypothesis cannot be deduced from the axioms of set theory. Like Gödel's theorem on the incompleteness of arithmetic, this fundamental result has general scientific and philosophical significance.

Cohen, a mathematician with wide-ranging interests, gave a paper at the 1962 congress in Stockholm devoted to harmonic analysis. Nevertheless, his principal achievement dealt with the continuum.

This survey is perforce brief and superficial, but the prize-winning papers bring out an impressive picture of mathematical progress over the last 50 years.

Tracing the fate of the Fields medalists makes one realize that the idea of the founder of the prize was exceptionally propitious. Nearly all winners of the prize are alive at present. The first two winners, J. Douglas and L. Ahlfors, passed away in 1965 and 1996, respectively. Many of them have continued to obtain significant results

[18] The 300-year history of Fermat's last theorem has apparently ended with a successful proof found by A. Wiles (Princeton University) [RS]. [Fal], [Po], [Wil].

even after receiving the award and are recognized authorities in their branches of mathematics. Many medalists have changed areas of research, but have achieved important results even in branches new to them. For example, Thom founded catastrophe theory, which has found important applications in mechanics, physics, ecology, and more. S.P. Novikov took up the study of the theory of nonlinear equations and the general theory of relativity. Smale has worked in economics and computational mathematics.

Recent years have seen other prominent international mathematical prizes appear. It is difficult as yet to pronounce on their longevity or compare them with the Fields prize. The most significant, the Wolf prize, is based on different principles [Za]. It crowns the career of great mathematicians (such at any rate is the appearance of the list of winners). Among the winners of the Wolf prize are Selberg, Ahlfors, Kodaira, Milnor, Hörmander, and Thompson. Whether or not the Fields medal can be compared with the Nobel Prize, Fields' happy idea of awarding it to the young has been crowned with complete success.

Appendix 1

The 1990 Fields Medalists

New Fields medalists were named at the Kyoto congress in August 1990. They were Drinfel'd, Jones, S. Mori, and Witten. All are well known in the mathematical world and highly deserve the honor. The only unprecedented happening was that the Fields prize was awarded for the first time to a person formally educated in physics and whose style of writing was physical, namely, E. Witten. In honoring him the mathematical community was recognizing the exceptional importance of the penetration of physical ideas and methods into modern mathematics. Recent papers of medalists Jones and Drinfel'd also concern to a degree mathematical physics, or, from a different point of view, physical mathematics. In light of the close connection among many results of these three researchers, I shall begin an analysis of the achievements of the 1990 medalists with Japanese mathematician Mori, whose work is somewhat peripheral to that of the others.

Shigefumi Mori. Mori, who specializes in algebraic geometry, continues the tradition of the distinguished Japanese school of algebraic geometers, which has already yielded two Fields medalists, Kodaira and Hironaka. Mori's most brilliant achievements are associated with the problem of classifying complex algebraic varieties of dimension ≥ 3.

In order to present Mori's results as concisely as possible, I shall make an excursion into the theory of algebraic varieties of low dimension. An algebraic variety is defined by a system of algebraic equations of the form

$$\begin{aligned} f_1(x_1, \dots, x_n) &= 0, \\ \dots\dots&\dots\dots \\ f_s(x_1, \dots, x_n) &= 0, \end{aligned}$$

where the f_i are polynomials in n variables. If the f_i are homogeneous polynomials, the variety can be regarded as a submanifold of projective space (CP^n, RP^n). In this case they are called projective algebraic varieties. Depending on the coefficient field F to which the coordinates of the points $(x_1, \dots, x_n)$ belong, one speaks of a real, complex, rational, or finite-characteristic algebraic variety.

Classifying one-dimensional complex algebraic curves is equivalent to the problem of classifying compact Riemann surfaces. The corresponding (unique) discrete topological invariant is the genus g—the "number of handles" of the Riemann surface. There are also continuous invariants (moduli) of the surface defined by integrals of holomorphic forms over cycles of the Riemann surface.

Curiously, the study of general properties of algebraic curves leads to the following division into classes depending on the genus of the curve g: 1) curves with $g = 0$; 2) curves with $g = 1$; and 3) curves with $g \geq 2$. The geometric analogues are respectively a sphere, a torus, and a Riemann surface of genus $g \geq 2$. With striking consistency, the general algebraic and arithmetic theorems for algebraic curves divide into precisely these three classes. The proof of Mordell's conjecture on the number of rational points on algebraic curves is the latest example of this type.

The classification of two-dimensional complex algebraic surfaces is incomparably more difficult. Here a significantly larger number of possibilities exist, and the problem is not yet fully solved. A certain method of classification exists, however.

In contrast to algebraic curves, by no means every compact complex surface is algebraic. Examples of such surfaces can be obtained from two-dimensional complex tori with a specially chosen lattice. Nevertheless methods of complex analysis are useful even in this problem.

Studying the behavior of holomorphic forms on algebraic varieties, Kodaira introduced an invariant that allowed him to distinguish two-dimensional algebraic surfaces in the spirit of algebraic curves. This invariant, called the Kodaira dimension, is connected with the multiple genera of the surface determined by the behavior of holomorphic forms of higher dimensions.

The Kodaira dimension facilitates a classification of surfaces analogous to that for curves, but the classification is far from complete. Not having the opportunity to give details of this beautiful classification, I note here also that the principle of "Neanderthal arithmetic" holds: 0, 1, many. The Kodaira classification includes surfaces of general type with $\mu = 2$ and special classes with $\mu = -1, 0, 1$. From the 1880s to the 1930s the Italian school of algebraic geometry obtained most of the results on the classification of complex two-dimensional surfaces. This school included

L. Cremona, E. Bertini, F. Enriques, G. Castelnuovo, and F. Severi [Di4].

At the beginning of this century, Enriques described all surfaces of a special type. But, putting the virtuoso analysis of the Italian geometers on a sound algebraic foundation required another 50 years of intensive development of algebra, topology, and algebraic geometry. R. Dedekind, H. Weber, Kummer, L. Kronecker, Hilbert, M. and E. Noether, and others laid the foundation for this work.

Despite the progress made in algebraic geometry in the postwar years, mostly related to introducing such new concepts as the theory of sheaves, schemes, the K-functor, and the like, the advance in more classical branches of algebraic geometry was slower.

The problem of a reasonable classification of multidimensional algebraic varieties conforming to the theory of algebraic curves or the Enriques-Kodaira theory of two-dimensional surfaces seemed intractable. For that reason Mori's papers in the 1970s and 1980s were unexpected. By introducing many new ideas he extended the Kodaira classification to projective algebraic varieties of dimension three. Greatly simplified, the method of Mori's classification can be thought of as the following sequence of operations.

Let X be a projective algebraic variety. I introduce the concept of the cone of X, denoted $C(X)$, i.e., the set of positive linear combinations of homology classes $H_2(X, R)$ of curves on X. The set $C(X)$ forms a cone in $H_2(X, T)$. Mori proved that a special basis consisting of smooth rational curves—the extremal rays—generates the negative intersection of this cone with a one-dimensional Chern class of the canonical class of X. Carrying out a contraction along the extremal rays, using Kodaira surgery, Mori derived all canonical forms of projective three-dimensional varieties. What is striking is that in the proof of the fundamental theorem he passed from the complex field to a field of finite characteristic. No proof of this theorem yet exists in the context of pure complex algebraic geometry. The methods developed by Mori have various applications and have already evoked many interesting results in multidimensional algebraic geometry. His ideas are highly significant for the entire foundation of this discipline.

One corollary of his work is a new, transparent proof of the Enriques-Kodaira classification theorem. The study of three-dimensional algebraic varieties with singularities is another topic. Mori's papers are a splendid combination of the algebraic-geometric structures of the Grothendieck era with the geometric intuition and analytic virtuosity of mathematicians of the golden age of algebraic geometry.

Let me complete this brief story of Mori's work with a remark on possible physical applications of his results. Modern theoretical physics, especially such branches as string theory and conformal field theory are absorbing the latest mathematical achievements with striking speed. Quite unexpected applications are now possible, but I mention only two areas right now. The first is the classification of n-dimensional (in particular three-dimensional) complex manifolds, a problem which arises in compactifying additional degrees of freedom in the theory of strings. The second problem is finding the connection between completely integrable dynamical systems and moduli of algebraic varieties.

Vladimir Gershonovich Drinfel'd. Continuing to violate alphabetical order and adhering rather to the intrinsic logic of narration, I now survey the papers of Drinfel'd, an algebraic geometer by education and a student of Manin. His first series of papers, which earned him international fame, deal with the solution of a critical problem in noncommutative class field theory—the proof of R. Langlands' conjecture for the group $GL(2)$ defined over a function field [Dr1]. Noncommutative class field theory, a central theme of modern algebraic number theory, naturally generalizes classical class field theory—the theory of abelian extensions of global and local fields.

Global fields include number fields, which can be regarded as finite extensions of the rational numbers, and function fields, i.e., finite extensions of the rational functions of one variable over a finite field. A local field is the completion of a global field, and an abelian extension is an extension with an abelian Galois group. Abelian class field theory basically seeks to describe an abelian subgroup of the Galois group $\mathrm{Gal}\,(K^s/K)$, where K is a local or global field and K^s the separable closure of it. Efforts of many generations of outstanding mathematicians of the nineteenth and twentieth centuries, among them E. Artin, Dedekind, Gauss, Hilbert, Kronecker, Kummer, J.L. Lagrange, T. Takagi, and Weil, solved this problem.

By understanding the first step taken by Lagrange and Kummer, who described the cyclic extensions of a field K, the reader can appreciate the full power of this theory.

The Lagrange-Kummer Theorem. *If a field K contains a primitive nth root of unity γ, then every cyclic field of degree n over K is generated by $\sqrt[n]{\delta}$ for some δ.*

The principal problems of number theory are connected to some extent with class field theory, for example, study of the distribution of zeros of the zeta-function over algebraic fields. The most accessible introduction to classical theory is Weyl's book [We].

After this introduction, it is easy to imagine the difficulties posed in the transition to the noncommutative case—the description of the full Galois group $\mathrm{Gal}\,(K^s/K)$. Langlands asserted that describing the group $\mathrm{Gal}\,(K^s/K)$ reduces to studying the set of finite-dimensional representations of the Galois group $\mathrm{Gal}\,(K^s/K)$. In more precise language the question is whether a one-to-one correspondence exists between the set of irreducible n-dimensional representations of the group $\mathrm{Gal}\,(K^s/K)$ and the set of automorphic representations of the group $GL(n, A)$. Here A is the ring of adeles—the set of numbers of the form $(a_\infty, a_2, \ldots, a_p)$, where a_∞ is a real number and a_p are p-adic numbers. Langlands' conjecture is a highly nontrivial. All of classical class field theory corresponds to the case $n = 1$.

The proof of this conjecture, aside from independent interest, leads to a solution of several fundamental problems of number theory related to the analytic properties of zeta-functions of algebraic varieties. Clearly every advance toward its proof is of exceptional interest.

Drinfel'd's first fully proved Langlands' conjecture for the case that follows the classical case ($n = 1$), namely the group $SL(2)$ over a function field.

Here we again encounter a striking fact that has been ubiquitous in the theory of numbers starting with the Riemann conjecture on the zeros of the zeta-function and ending with the Mordell conjecture. Proof of the theorem has been obtained for function fields. For number fields, even in the case $GL(2)$, Langlands' conjecture remains unproved, and even stating it involves certain difficulties.

Drinfel'd not only obtained results having various applications in algebraic geometry, but also the method of proof. He introduced new algebraic-geometric structures, the F-sheaves, that find wide application across algebraic geometry, for example, in the theory of algebraic surfaces.

Drinfel'd's interests are not limited to algebraic geometry. Modern physical problems, where mathematicians may find applications of their methods and acquire new formulations of purely mathematical problems, did not escape Drinfel'd's attention. I have already noted the work of Atiyah, Drinfel'd, Manin, and Hitchin on the algebraic-geometric classification of instantons. This work made Drinfel'd widely known among theoretical physicists. Papers of Drinfel'd and A.A. Belavin on the construction of the solutions of the Yang-Baxter equations ("triangles") in conformal field theory and his joint paper with V.V. Sokolov on nonlinear equations followed. They showed that for each infinite-dimensional Kac-Moody algebra it is possible to construct a system of evolution equations linked to a two-dimensional integrable system (a Toda chain) just as the Korteweg-de Vries equation is connected with the sine-Gordon equation [DS]. The proof that the equations of two-dimensional gravitation are integrable has recently drawn additional attention to this paper.

Finally, I come to the last series of papers by Drinfel'd, which has brought him the most fame, on quantum groups [Dr2]. The object itself, quantum groups (choosing a propitious name is half the job) arose earlier under different names—Hopf algebras, ring groups—and in relation to other problems. Somewhat inaccurately, a quantum group is an algebra on which two kinds of multiplication are defined: multiplication, which maps the vector space into the algebra, and comultiplication, or multiplication in the dual space.

H. Hopf first studied algebras of this type in algebraic topology in constructing the theory of cohomology of groups. Milnor and J. Moore systematically studied Hopf algebras from the point of view of topology. In the 1960s G.I. Katz introduced another class of quantum groups—ring groups—in generalizing the concept of duality on noncommutative groups. Here the analogues of quantum groups are rings of continuous functions on a group. Hopf algebras appeared, also in algebraic field theory, in the laws of superselection. Nevertheless these profound results seemed artificial and to have an insignificant field of applications.

In the last decade the situation has changed radically with the discovery of a powerful new method of integrating the model equations of quantum field theory and statistical physics—the quantum inverse scattering method. The new method is an amalgam of ideas from the classical inverse problem and specific devices gleaned from physics—Bethe substitutions and the Yang-Baxter equations.

The main object of this theory is the so-called R-matrix. This matrix makes it possible to compute the transfer-matrix of the corresponding equation and find the spectrum of the model. The algebra of R-matrices was to be a very nontrivial mathematical object. In the analysis of R-matrices of solutions, Drinfel'd discovered their close connection with quantum groups.

Together with quantum groups he introduced Poisson-Lie groups. His definition of a Poisson-Lie group G included a definition of Poisson brackets on the function space $F(G)$ and the coproduct $\Delta : F(G) \to F(G) \otimes F(G)$.

With this definition one can introduce the concept of the quantized Poisson algebra as a deformation depending on the parameter $\hbar$ (where $\hbar$ has a physical interpretation as Planck's constant). This last property justifies the term quantum group.

Quantum groups were a concept whose time had come. Simultaneously with the papers of Drinfel'd, works of M. Jimbo and S. Woronowicz appeared that gave other definitions and applications of quantum groups. The Leningrad school of mathematical physics founded by L.D. Faddeev [Fa] arrived at first-rate results in the theory of quantum groups. Faddeev and his students have been leaders in the construction of a general method of integrating quantum systems: the construction of the quantum inverse scattering problem. No wonder such active participants in the group as V.E. Korepin, N.Yu. Reshetikhin, M.A. Semenov-Tyanshanskii, S.E. Sklyanin, L.A. Takhtadjan, and others have taken up the development of the theory of quantum groups. The close contacts with topology (V.G. Turaev, Viro) lead to unexpected ties with another recent remarkable discovery—the Jones polynomials. I shall say more about this later when I discuss another winner of the Fields prize.

Vaughan Jones. In 1990, for the first time, the Fields prize honored a mathematician from the southern hemisphere. Jones was born in New Zealand, graduated from the University of Geneva, and is now professor at the University of California at Berkeley (the nightmare of Soviet[19] bureaucrats—the continuing "brain drain").

A student of Swiss topologist A. Haefliger, Jones specialized in one branch of functional analysis, the theory of factors. In connection with his research in the von Neumann theory of factors he made a remarkable discovery—the construction of a new type of polynomial invariants of knots. In the main portion of this book I have already mentioned the Jones invariants, and so I shall try here to give a clearer idea of the structure of the Jones polynomials proper.

I begin with a parallel description of two systems. The first system is physical. Consider the motion of three points on a line, all moving with constant velocity. Assuming that only pairwise collisions are possible and that the particles exchange only their internal degrees of freedom in a collision and maintain the trajectories of their motion, we obtain the following equation for the scattering matrices of three-particle scattering:

$$R_1 R_2 R_1 = R_2 R_1 R_2. \tag{13}$$

Here $R = (R_{ki}^{(lj)})$ are the scattering amplitudes; i and j are the final states; and k and l the initial states of the system of particles $(i, j, k, l = 1, 2, 3)$. Here the indices 1 or 2 indicate that the matrices R act on the particles (1 2) and (2 3) respectively. Equation (13) is the simplest example of the Yang-Baxter equations.

[19] Now "former" Soviet bureaucrats. Human psychology does not change as rapidly as political regimes.

I now turn to a mathematical model. With each knot one can associate the group braid B. The simplest nontrivial braid consists of three threads a_1, a_2, and a_3. If we identify isotopic sets of threads, the set of braids becomes a group B_3. There exists a homomorphism of the group B_3 into the permutation group S_3. The elementary generators of the group σ_i ($i = 1, 2$) satisfy relations analogous to (13). This fact holds also for an arbitrary group B_n.

By associating the matrix R_n with the element σ_n, we obtain a matrix representation of the group B. Computing the generating function for this representation gives the Jones knot invariants.

This construction makes clear the role of the Yang-Baxter algebra for finding a system of invariants. Immediately after the work of Jones two directions of research came to the forefront. One is connected with topology. In this area researchers obtained new classes of invariants: entire series of new polynomials, for example polynomials in two variables, Kaufman polynomials, and the like.

In contrast to the Alexander polynomials, the Jones polynomials have the remarkable property of chirality, i.e., they distinguish a knot from its mirror image. For the simplest nontrivial knot, the trefoil for example, the Alexander polynomial $A(t)$ is $t^2 - t + 1$, while the Jones polynomial is $V = -t^4 + t^3 + t$. Computing the Jones polynomial solves in some cases the old problem of distinguishing a knot from its mirror image.

Connections of Jones structures with physical problems are of equal interest. The original construction was based on the study of factors that arise in exactly solvable models of statistical physics. A direct link was soon found between the method of the quantum inverse problem (R-matrices) and polynomials of Jones type in papers of Jones and Turaev. By choosing a suitable Yang-Baxter equation, it is possible to construct other polynomials of Jones type. Most recently knot invariants have been obtained by using the ideas of quantum group theory.

For all their remarkable properties the Jones polynomials had certain shortcomings. It remained unclear how the Jones polynomials relate to the known topological constructions of knot invariants: the fundamental group of the complement of the knot, linking coefficients, Milnor numbers, Seifert manifolds, and so forth.

Moscow mathematician V. Vasil'ev recently proposed an unexpected approach to the problem of constructing a complete system of knot invariants. Vasil'ev's work is giving a new look at this whole body of questions. [Va]

The Vasil'ev construction of knot invariants is based on ideas of singularity theory and schematically looks as follows. Consider the set of mappings $\mathbb{S}^1 \to \mathbb{S}^3$ having singularities or self-intersection. This set is called the *discriminant* and forms a special hypersurface in the space of all mappings. Its nonsingular points correspond to mapping with one point of transversal self-intersection, and its singularities to mappings having derivatives with zeros, or nontransversal or multiple self-intersections. By use of a discriminant any numerical knot invariant of isotopic type can be given. To be specific, to each nonsingular piece of the discriminant, i.e., connected component of the set of its nonsingular points, one must ascribe an index—the difference of values of the invariant for the nearby knots separated by this piece. This set of indices is not arbitrary and must satisfy a homology condition if the invariant is to be well defined: the sum of the components taken with certain coefficients is homologous to zero in the space of mappings $\mathbb{S}^1 \to \mathbb{S}^3$. A more precise definition requires introducing the class of noncompact knots, i.e., nonsingular imbeddings $\mathbb{R}^1 \to \mathbb{R}^3$ of infinity tending to a fixed linearly imbedded line $\mathbb{R}^1$. I shall denote the space of all smooth mappings, including singular mappings, by $\mathcal{K}$. This space is homotopically trivial. I denote the discriminant of this space by $\mathcal{D}$. The connected components of the space of noncompact knots $\mathcal{K} \setminus \mathcal{D}$ is in one-to-one correspondence with the regularly homotopic classes of ordinary knots $\mathbb{S}^1 \to \mathbb{S}^3$. The Vasil'ev invariants correspond to the zero-dimensional cohomology group $H^0(\mathcal{K} \setminus \mathcal{D})$. The group $H^0(\mathcal{K} \setminus \mathcal{D})$ together with $H^i(\mathcal{K} \setminus \mathcal{D})$, $i > 0$ is computed using the spectral sequence whose filtration is determined by the types and multiplicities of singularities of the discriminant surface.

The fundamental question of knot theory—whether there exists a complete system of invariants—reduces in the Vasil'ev theory to determining the convergence of the spectral sequence. Currently there is not a complete answer, but the Vasil'ev theory appears to be the most realistic route to solving this problem. Even the preliminary results and connections discovered with other, remote branches of mathematics, show the exceptional importance of this result. I note only a few of these results, which have all been obtained very recently.

J. Birman and X.S. Lin first showed how to obtain the Jones polynomials from the Vasil'ev invariants [Bi] which is only the initial, but important step, on the road to understanding the Vasil'ev and Jones invariants within the classical topological technique. M. Kontsevich, D. Bar-

Natan, and Witten are developing an intriguing approach to the Vasil'ev invariants, based on topological field theory. In particular, Kontsevich has found integral representations like the Gauss formula for Vasil'ev invariants, and Bar-Natan and Witten have devised a perturbation theory for computing the same invariants, using Feynman integrals for the Chern-Simons action. The Vasil'ev technique makes possible the construction of analogous invariants for imbeddings of multi-dimensional knots.

Edward Witten. The prior results indicate a discovery of exceptional importance, while the unexpected connections with other fundamental problems of mathematical physics, sometimes encountered, suggest that we are only at the beginning of the road. Witten's papers best confirm this fact. Witten graduated from Harvard University having majored in physics. His first paper, which made him famous, was the construction of an n-instanton solution of the Yang-Mills equation. It showed a physicist who had an impressive mastery of modern mathematical machinery. In about four years he wrote a series of papers having both purely physical interest and great mathematical value. The fundamental paper was "Supersymmetry and Morse theory," which has already been mentioned in discussing the Atiyah-Singer index theorem. The principal idea, the construction for each classical differential operator on a manifold of a certain fermion operator of the classical differential operator accompanying the quantum-mechanical supersymmetric system, has been exceptionally productive.

Witten is a brilliant representative of the new wave of physicists who are redefining the face of modern theoretical physics, turning it more toward mathematics. While other physicists—it suffices to mention A. Polyakov, A. Zamolodchikov, and G. 't Hooft—arrive at new or nonstandard mathematical applications (the theory of monopoles, instantons, conformal field theory) in solving a physical problem, the opposite trait is characteristic of Witten, especially in his later papers: He applies physical ideas to construct mathematical structures. Here are two examples. The first involves a new approach to the construction of Jones polynomials.

Let M^3 be a three-dimensional closed manifold and L a set of linked circles l_i, situated in it. With each manifold M^3 we connect a certain topological field-theoretic model defined by the Lagrangian

$$\mathfrak{L} = k \int_{M^3} \mathrm{Tr}\,(A \wedge dA + (2/3) A \wedge A \wedge A).$$

Here A is the connection, or field intensity, generated by the bundle over M^3 with gauge group G. With the system of curves l_i one can associate a functional of the connection A.

$$W_R(l_i) = \mathrm{Tr}_R P \exp \int_{l_i} A_i\, dx^i,$$

where R is an irreducible representation of the group G, P is the normal ordering necessary for defining exp for a noncommutative group G.

Now consider the Feynman path integral over:

$$Z = \int DA \exp(i\mathfrak{L}) \Pi W_{R_j}(l_j).$$

DA is the Feynman measure.

Proposition. *The correlation function Z defines Jones-type invariants for the link L.*

One corollary is the possibility of defining the Jones invariants for linkages lying in any compact manifold M^3, not only in the sphere $\mathbb{S}^3$. Although Witten's chief paper does not meet mathematical standards of rigorous proof, the large number of brilliant ideas, conjectures, and results it contained more than atoned for this sin.

The second example, taken from a 1992 paper by Witten, relates to a major discovery in theoretical physics—the exact solutions of the two-dimensional gravity equations. This significant result seems particularly interesting in its method. In essence Witten arrived at the first major result in field theory by combining analytic computations and computer experiments. This achievement is based on the method of dynamic triangulations of manifolds, which A. Migdal and his school had developed over many years [Mig]. With his characteristic energy Witten took up this problem. He conjectured that a connection exists between the problem of the moduli of Riemann surfaces and topological gravity. He found a representation for the intersection numbers of the moduli space of a punctured Riemann surface of genus g in terms of solutions of the generalized Korteweg-de Vries equation, the basic object of the theory of completely integrable systems. Kontsevich [Wi3] has obtained a strict proof in some special cases.

Witten's ideas in topological field theory allow us to sense hidden connections between such beautiful results (at first sight so remote from one another) as the theory of Jones polynomials, Donaldson and Floer invariants of three- and four-dimensional manifolds, completely integrable equations of conformal field theory, two-dimensional statistical systems, and many others.

In recognizing Witten, the Fields committee set a precedent that seems to be in perfect accord with the spirit of this prize. By choosing Witten along with three "regular" mathematicians—Drinfel'd, Jones, and Mori—the Fields committee was reminding the mathematical community of the range of remarkable discoveries that had been made in modern mathematics and what penetrating talents are at our disposal. The Fields prize entered its second half-century with confidence.

Appendix 2

The 1994 Fields Medalists

Zürich had the honor of hosting the 1994 International Congress of Mathematicians. In this city the history of mathematical congresses had begun in 1897,[20] and the Fields prize was established there in 1932. Considering that up to the present the International Congress has not been held even twice in any other city, the role of Zürich in the life of the worldwide mathematical community is truly exceptional.

J. Bourgain, P.-L. Lions, J.-C. Yoccoz, and E. Zelmanov are the new Fields medalists. These mathematicians had obtained significant results in algebra, harmonic analysis, the theory of dynamical systems, and partial differential equations. The 1994 decisions of the Fields committee bespeak a reaction to the tendency of the preceding few congresses, where topology and algebraic geometry predominated.

In this brief survey of the achievements of the Fields winners I shall follow the classifications of the congress and

---

[20] The 1893 congress in Chicago, organized by Felix Klein, was not international, despite being called so. Only four foreign scholars took part in the congress, which was part of the World's Fair celebrating the 400th anniversary of Columbus' discovery of America.

begin with the classical section and the oldest branch of mathematics—algebra.

Efim Zelmanov. Zelmanov received the Fields medal for solving the restricted Burnside problem. This result capped off an extended period in group theory.

In 1902 British mathematician W. Burnside stated the following problem. Consider a group $B(m,n)$ with a finite number of generators $b_1, \ldots, b_m$, all of whose elements have finite order n: $g^n = 1$, $g \in B$ (such groups are said to be *periodic* of degree n). Is the group B finite? A weaker statement, which came to be known as the generalized Burnside problem, reduces to the question of the finiteness of the group B under the assumption that it is periodic, without requiring a universal degree n for all elements.

The solution of the Burnside problem and its generalized statement was exceptionally difficult but fruitful. The solution took over 60 years and required new, powerful methods, not only in abstract group theory, but also the application of ideas from other areas of algebra, like the theory of Lie algebras.

In 1964 Moscow mathematician E. Golod solved the generalized Burnside problem. He constructed an example of a finitely generated infinite group all of whose elements have finite order, but for which the orders are not uniformly bounded. His result is based on reducing the generalized Burnside problem to a certain problem in the theory of Lie algebras, where analogous ideas made it possible to construct a counterexample to another interesting conjecture: the finiteness of a tower of fields of classes [Gol].

About the same time an assault on the classical Burnside problem began. P.S. Novikov (the father of S.P. Novikov) took a fundamental step in 1959, but the complete proof was not achieved until 1968 in a joint paper of P.S. Novikov and S.I. Adyan. They constructed an example of an infinite periodic group having odd degree $n \geq 4381$. The proof of this result is among the most difficult in modern mathematics. It occupies more than 300 journal pages and is based on a complicated induction. In recent years Adyan, A.Yu. Ol'shanskii, I.G. Lisenok, and others, have succeeded in simplifying the original proof, lowering the degree n to 115.

These results make it all the more remarkable that the restricted Burnside problem has a positive answer. The explicit statement and terminology are due to W. Magnus (1950): *Is the number of m-generated finite groups of degree n finite?* More precisely, the question reduces to the existence of a maximal finite m-generated group $B_0(m,n)$ of degree n: $B_0(m,n) = B(m,n)/H$, where H is the intersection of all normal subgroups of finite index in the group $B(m,n)$. Magnus reduced this problem for a prime exponent $n = p$ to the question of the nilpotence of finitely generated Lie algebras with the Engel identity. Subsequently I. Sanov generalized this result to $n = p^\alpha$ and P. Hall and G. Higman proved that the case of general n reduces to $n = p^\alpha$.

A. Kostrikin [Kos] took the next step. He proposed a proof of the restricted Burnside problem for a prime exponent $n = p$. Unfortunately, it contained serious gaps and was completed only 20 years later. For $n = p^\alpha$ the proof becomes significantly more complicated, since the corresponding algebra L is no longer an algebra over the ring $\mathbb{Z}_p$, and satisfies a weaker condition than the Engel identity, the so-called linearized Engel condition. Nevertheless by using highly nontrivial constructions, including Jordan algebras and supersymmetric Lie algebras, Zelmanov proved that the algebra L is nilpotent. It follows that the restricted Burnside problem has a positive answer for exponents $n = p^k$, including the case $p = 2$ [Ze1, Ze2].

For an arbitrary n the proof is achieved by reduction to the theorem on the classification of finite simple groups. It would be extremely desirable to have a direct proof of the general case.

Zelmanov harmoniously combines a mastery of virtuoso techniques with the use of new ideas and general algebraic constructions. One important ingredient in the proof of the restricted Burnside problem is Jordan algebras. For many years Zelmanov studied the theory of Jordan algebras. He made significant advances in the classification of finite-dimensional and infinite-dimensional Jordan algebras [Ze3].

The attempt to extend Zelmanov's results to compact and so-called profinite groups gave rise to yet another circle of problems. Zelmanov proved an analogue of the generalized Burnside problem for periodic compact groups. A corollary of this result is a beautiful theorem: *Every infinite compact group contains an infinite abelian subgroup* [Ze4].

Despite these beautiful results connected with Burnside's problem, many unsolved problems remain. For example, no lower bound has been found on the degree n in the Novikov-Adyan theorem.

But connections with other branches of mathematics may be yet more important. The connection with the hyperbolic Gromov groups is promising [Grom]. Undoubtedly the theory of Jordan algebras, which arose in connection with problems of quantum mechanics in the work of P. Jordan, von Neumann, and E. Wigner and was so brilliantly advanced in the work of Zelmanov, will find substantial physical applications.

Jean Bourgain. Bourgain, a mathematician with wide interests, is the author of major results in Banach spaces, harmonic analysis, convex bodies, ergodic theory, and nonlinear equations. Characteristic of his work is his virtuoso mastery of analytic technique and the skillful construction of unexpected examples, some in well-traveled and traditional areas.

Let us begin with the theory of Banach spaces, a classical branch of linear functional analysis. Banach spaces, discovered by the Polish mathematician Stefan Banach in the early 1920s, were intensely studied in the period just before the Second World War. The Polish school, which included some outstanding mathematicians, such as H. Steinhaus, J. Schauder, S. Mazur, and others, made an especially large contribution to the theory of Banach spaces. Unfortunately the war interrupted their fertile investigations. After the war, interest in Banach spaces declined sharply. A feeling had arisen that, outside the difficult unsolved problems left by classical authors, little prospect existed of any particularly interesting results, much less of applications to other branches of mathematics. The change in this point of view, which occurred in the mid-1950s, is largely bound to the new ideas and methods propounded by Grothendieck and Schwartz. The "great French revolution," which reawakened interest in this area of mathematics, was the solution of several of the most difficult problems in the theory of Banach spaces.

In 1973 Per Enflo's negative solution of the famous problem of the existence of a basis caused a sensation. One important corollary of this work was the identification of a large class of Banach spaces having special (Schauder) bases.

Such spaces have the approximation property, i.e., every compact operator from any Banach space into the given Banach space B can be approximated by operators of finite rank. Banach spaces of this type are infinite-dimensional, yet many of their characteristics are close to finite-dimensional spaces. Nevertheless, even general Banach spaces have a number of remarkable properties. Well-known examples of Banach spaces are the spaces $L^p(S)$, $1 \leq p < \infty$, L^∞, $l_p(S)$, $C(S) = \{c(S)\}$, where $C(S)$ is the space of bounded scalar-valued functions defined on the set S. From the topological point of view the Banach spaces of simplest structure are the Hilbert spaces $L^2(S)$ and l_2. In particular all separable Hilbert spaces are isomorphic. For general Banach spaces this assertion no longer holds. Examples constructed by Bourgain show the complicated phenomena that arise in the study of arbitrary Banach spaces. Here are two examples from the many constructed by Bourgain: 1) There exists a subspace $V \subset L^1$ isomorphic to l_1 but having no complementary subspace $V^\perp$ such that $V^\perp \oplus V = L^1$; 2) in l_1 there exist uncomplemented subspaces Y isomorphic to l_1.

Recently these results have aroused the interest of specialists in Banach algebras in the attempt at a homological classification of modules over rings. These ideas also have earlier work of Grothendieck as a source.

Important Banach spaces that arise in several branches of analysis are the spaces of analytic functions defined in a domain of the complex plane $A(D)$, where $D \subset \mathbb{C}^n$. A fundamental question asks whether the isomorphism of two spaces $A(D_1)$ and $A(D_2)$ is determined by the dimension and geometry of the domains D_1 and D_2. Among the profound results obtained in this area was G.M. Henkin's proof that the spaces $A(D_k)$ and $A(D_1^n)$ are nonisomorphic for any natural numbers $n \geq 2$, $k \geq 1$. Here D_k is the open unit ball in $\mathbb{C}^k$ and D_1^n is the unit polycylinder in $\mathbb{C}^n$: $|z_i| < 1$, $(i = 1, \ldots, n)$. Bourgain solved the analogous problem for the Hardy spaces $H^1(D)$, which are subspaces of the corresponding $A(D)$. He proved that the spaces $H^1(D_m)$ and $H^1(D_n)$ are nonisomorphic, where D_m and D_n are polydisks of the corresponding dimensions. The study of isomorphism required a complicated analytic technique, in particular the theory of integral representations in complex domains. Bourgain also solved the difficult basis problem in the space $A(B)$, where B is the unit ball in $\mathbb{C}^n$. Previously it had been known that an unconditional basis exists in $A(B)$. Bourgain showed that there is no Schauder basis in $A(B)$. This result shows that delicate properties of Banach spaces arise even in the most natural examples. The similarity to finite-dimensional vector spaces makes it possible to develop the beautiful geometric theory of Banach spaces. Remarkable applications have opened up in the theory of approximations of functions, harmonic measures, entropy properties of function spaces, and so forth. Bourgain, jointly with V. Milman, obtained the following result: *Let X be an n-dimensional real normed space with unit ball B and ε an ellipsoid of maximal volume contained in B. Then the following estimate holds:*

$$\left(\frac{\mathrm{Vol}_n B}{\mathrm{Vol}_n \varepsilon}\right) < kC_2(X) \log^4 C_2(X),$$

where $C_2(X)$ is defined for every finite-dimensional space X.

This result is valid for a large class of Banach spaces, for example L^p, $p \geq 2$, and has many applications in the theory of Kolmogorov diameters.

Bourgain's work on harmonic analysis in real spaces generated great resonance. His analytic talent fully revealed itself in the study of multi-dimensional Fourier transforms and generalized oscillatory integrals. To obtain precise estimates of the spherical means in the metric of L^p he applied new geometric constructions, including a multi-dimensional generalization of the Besicovitch-Kakeya sets.

Not having space to discuss all of Bourgain's work, much of which is devoted to solving long-standing problems by relying on complicated analytic machinery, I shall mention only one result, which can be stated simply. The ergodic theorem proved by Bourgain is a beautiful generalization of Birkhoff's well-known theorem.

Let T be a measure-preserving ergodic transformation of the space Ω and $f \in L^r(\Omega, \mu)$, where $r > 1$ and μ is a probability measure on Ω. Let $P(x)$ be a polynomial with integer coefficients. Consider the mean:

$$A_N f = \frac{1}{N} \sum_{1 \leq n \leq N} T^{p(n)} f.$$

Bourgain's Theorem. *The quantity $A_N f$ converges almost everywhere on Ω.*

This theorem applies to a large class of arithmetic sets and has applications in number theory and harmonic analysis.

In recent years study of the existence of global solutions of nonlinear evolution equations has entered Bourgain's interests, and he has applied the technique of harmonic analysis. A brief survey of Bourgain's achievements and a selected list of his papers appears in the article by J. Lindenstrauss in the *Notices of the American Mathematical Society* [Lin].

Pierre-Louis Lions. Lions concentrated on the theory of partial differential equations, an area of mathematics closely connected with physics, mechanics, and control theory. An extraordinarily active mathematician, Lions has obtained profound results in several areas of differential equations.

I shall single out two series of papers of Lions, which have brought him great renown.

1. The theory of kinetic equations.
2. The theory of viscosity solutions.

The kinetic theory, which mathematically describes the most important physical media such as gases, plasma (ionized gas), and the like, is a complicated system of consistent nonlinear integro-differential equations. Pursuit of general methods of solving such equations, and clarification of conditions for uniqueness and smoothness of the solutions are very difficult mathematical problems.

A classical kinetic equation is the system

$$\frac{\partial f}{\partial t} + v\frac{\partial f}{\partial x} + \dot{p}\frac{\partial f}{\partial p} = \mathrm{CI}\, f, \tag{14}$$

where f is the distribution function of particles over the coordinates (x) and momenta (p), and $\mathrm{CI}\, f$ is the collision integral. The derivative $\dot{p}$ is determined by the force acting on the particle. Equation (2) contains as special cases the Boltzmann equation ($\dot{p} = 0$) and the equation that describes a collision-free plasma. In a series of papers, written jointly with R. Di Perna, Lions studied the Cauchy problem for Equation (14). He distinguished a class of global solutions, the so-called *renormalized* solutions, for which a priori estimates derived from conservation laws are valid. In the case when $\mathrm{CI}\, f = 0$, Lions and B. Perthame proved the uniqueness and regularity of the corresponding solutions. The methods of proof are based on deriving delicate a priori estimates and require the application of nonstandard considerations from harmonic analysis. These methods have also been applied to other physically interesting problems: the Navier-Stokes equation, the equations of gas dynamics, and a number of others.

Since the classical work of E. Hopf, the method of viscosity solutions has been used in the theory of differential equations. Hopf's idea can be explained using the example of the simplest quasi-linear equation

$$u_t + uu_x = 0. \tag{15}$$

Consider the auxiliary equation

$$u_t + uu_x = \varepsilon u_{xx}, \tag{16}$$

with initial condition $u_{t=0} = u_0(x)$.

Passing to the limit over the vanishing viscosity $\varepsilon \to 0$ in the solution $u = u^\varepsilon(t, x)$ of Equation (16) gives the solution of Equation (15). A suitable justification of passage to the limit requires estimates of the solutions $u^{\varepsilon}(t,x)$ that are uniform with respect to ε for $u_0(x) \in L^1(\mathbb{R}^1) \cap L^\infty(\mathbb{R}^1)$. Equation (15) belongs to an important class of equations generated by the system of conservation laws

$$u_t + \varphi(u)_x = 0$$

with convex function $\varphi(u)$.

Lions broadened and generalized the theory of viscosity solutions. The foremost class of equations amenable to the method of viscosity solutions includes the Hamilton-Jacobi equations and the Bellman equation, which is fundamental to control theory.

A much larger field of applications arose after the work of R. Jensen [Je], who proved that the theory of viscosity solutions applies to second-order equations. This body of questions harmoniously encompassed wide application of nontraditional methods, mainly the wide application of the ideas of global functional analysis, and a significantly enlarged area of applications.

I have mentioned only two areas of Lions' work. Not having space to discuss his other interesting results, I refer the reader to the articles by Lions [Lio1], [Lio2], and the paper [CIL], which contains detailed information.

Jean-Christophe Yoccoz. Yoccoz pursued two rapidly developing areas of mathematics: the theory of dynamical systems and holomorphic dynamics. The two areas are closely intertwined and connected with the names of classical scholars of the French mathematical school: Poincaré, Fatou, and Julia. The fates of these two areas were different. Poincaré's work in celestial mechanics and his paper

"On curves defined by differential equations" laid the foundations of the modern theory of dynamical systems. The leading mathematicians of the twentieth century extended them. The papers of Fatou and Julia on the structure of the endomorphisms of complex sets, such as the Riemann sphere, remained a backwater in mathematics for almost 50 years. Only in the late 1960s was interest revived in connection with new horizons in the theory of dynamical systems, Sinai billiards, and the like. In the last decade studies in holomorphic dynamics have become especially popular. Such new and "well-forgotten old" concepts as fractals and strange attractors have now found a natural sphere of application.

In this more intuitive presentation of Yoccoz' achievements, I shall briefly state several basic problems in both dynamical systems and holomorphic dynamics and mention the results of his illustrious predecessors.

1. The problem of small denominators.

In his studies on the three-body problem Poincaré grappled with the difficult series of perturbation theory applied by astronomers G.W. Hill, H. Gyldén, A. Lindstedt, and others in studying the stability of the orbits of planets and asteroids.

The small denominators that arise in the analysis of these series, i.e., expressions of the form

$$\langle \lambda, q \rangle, \tag{17}$$

where $\lambda = \{\lambda_1, \ldots, \lambda_n\} \in \mathbb{C}^n$, $q = \{q_1, \ldots, q_n) \in \mathbb{Z}^n$, and $\langle \lambda, q \rangle$, the inner products occurring as denominators in the terms of the expansion

$$\sum a_{mn} \frac{e^{i\langle \lambda, q \rangle t}}{\langle \lambda, q \rangle}, \tag{18}$$

turn out to be anomalously small and make analysis of the asymptotic convergence of the series (18) difficult. Without this analysis operations with the series become unfeasible. The problem of small denominators also arises in other problems of the theory of ordinary differential equations, for example, in studying the behavior of the trajectories of conservative systems in a neighborhood of a point of equilibrium or a periodic solution. An effective method of solving this problem is to reduce the equation to a simpler form, i.e., to find a normal form. A typical example is the following problem: Suppose given the equation

$$\dot{x} = A(x), \quad x \in \mathbb{C}^n, \tag{19}$$

where $A(x)$ is a power series in x. Under which conditions can this equation can be made into a linear equation by a change of the variable x:

$$\dot{y} = Ay? \tag{20}$$

This problem has various formulations and generalizations, for example, reduction to the form (20) in the class of formal power series, which Poincaré himself studied. The question of the reduction using analytic functions is more complicated. It requires clarification of the convergence of the formal series. These problems formed part of the classical studies of Birkhoff, Siegel, and other prominent mathematicians. Despite all these efforts, the multi-dimensional problem of finding a normal form remains far from being completely solved. But in the case of two degrees of freedom it has been solved. The importance of the properties of arithmeticity of the eigenvalues of the matrix A was already cited in the early papers of Poincaré. Later Siegel proved the following remarkable theorem:

If the eigenvalues of the linear part of the matrix A at a singular point satisfy the condition

$$|\lambda_i - (m, \lambda)| > \frac{c}{m^\nu}, \quad \nu > 2,$$

then the field $A(x)$ is analytically equivalent to its linear part in a neighborhood of the singular point.

This theorem also holds in the multi-dimensional case. T. Cheery and A. Bryuno made it possible to weaken Siegel's condition when they proved the following theorem:

Let $\lambda = \lambda_1/\lambda_2$, and let λ have the continued-fraction expansion

$$\lambda = a_0 + \cfrac{1}{a_1 + \cfrac{1}{a_2 + \cdots}}.$$

Let q_k denote the denominator of the convergent $\frac{p_k}{q_k}$ of order k in the expansion of λ. Then the system (19) *can be reduced to the form* (20) *if*

$$\sum_{k=1}^{\infty} q_k^{-1} \ln q_{k+1} < \infty.$$

Although this result was reached from 1964 to 1965, it took another 20 years before Yoccoz proved the necessity of this condition. Thus, in the case of two degrees of freedom a complete criterion for analytic reducibility to normal form (20) exists.

Another result of Yoccoz has a long prehistory and also goes back to Poincaré, namely the study of conditionally periodic motions of Hamiltonian systems. Study of the motions of Hamiltonian systems even on an invariant torus ($\mathbb{T}^2$) with two degrees of freedom leads to profound theorems in the theory of dynamical systems.

Let us define a differential equation on the torus $\mathbb{T}^2$ in the form

$$\frac{dx}{dt} = F(x,y), \quad \frac{dy}{dt} = G(x,y),$$

where $F(x,y)$ and $G(x,y)$ are functions having a certain smoothness and periodic in x and y. Under which conditions on functions F and G can the system be reduced to the form

$$\frac{dx}{dt} = \lambda_1, \quad \frac{dy}{dt} = \lambda_2? \tag{21}$$

Thus the question is whether the system (21) admits conditionally periodic motions with particles λ_1 and λ_2. The solution of this problem and its generalizations occupied the entire twentieth century and led to many fundamental concepts in the modern theory of dynamical systems.

Here is a sketchy history of the subject. Poincaré took the first decisive step when he showed that the system (17) can be reduced to mapping a circle into itself. To this end he introduced a sweep function connected with Eq. (17). A sweep function is a mapping of a meridian of the torus onto itself taking each point of the meridian to the next point of intersection of the trajectory of the equation with the meridian. Thus the study of trajectories reduces to the study of homeomorphisms of a circle. The second concept, also introduced by Poincaré, is the winding number. In terms of homeomorphisms of the circle f it can be defined as $\lim\limits_{n\to\infty} \dfrac{f^n(x_0)}{n} = \rho$. This number is independent of the choice of the point x_0 on $\mathbb{S}^1$. The winding number of the homeomorphism $\mathbb{S}^1$ determines the number of revolutions of the vector field of Equations (17) and (18). For Equation (18), for example, we have $\rho = \lambda_1/\lambda_2$. This number is an important characteristic of the homeomorphism. The arithmetic properties of ρ determine the behavior of homeomorphisms. Thus if $\rho = p/q$ is rational, there exist periodic trajectories of period q that traverse the circle p times. In the general case the trajectories of almost all points are attracted to stable periodic trajectories. If ρ is irrational, then for sufficiently smooth diffeomorphisms T_f (for example of class C^2) the trajectory of any point is dense in the circle, and the diffeomorphism reduces to rotation by angle ρ. A. Denjoy found this result. It can be stated as the following relation:

$$T_g T_f T_g^{-1} = T_\rho,$$

where T_ρ is $\{x + \rho\}$, a rotation of the circle by angle ρ. Denjoy's theorem assumes nothing about the smoothness of the transformation T_g.

This problem is crucial in constructions in perturbation theory for conditionally periodic motions, since it is linked to the condition of conservation of invariant tori. In KAM theory, named for its founders Kolmogorov, Arnol'd, and J. Moser, this problem was solved for analytic and sufficiently smooth diffeomorphisms of a circle close to rotations. The corresponding results came to be known as local reduction theorems. The possibility of reducing T_f to a rotation was found to depend on the arithmetic properties of ρ, more precisely on the rate at which an irrational number ρ can be approximated by rational numbers. M. Hermann (Yoccoz' advisor) took the fundamental new step when he proved a global reduction theorem, except for the requirement of nearness to a rotation. Hermann showed that if the homeomorphism T_f has smoothness C^3, then for almost every winding number ρ a diffeomorphism T_f having winding number ρ is smoothly equivalent to a rotation through the angle ρ. In this theorem "almost every" means that the measure of any excluded set is zero. Yoccoz thereby greatly strengthened the result of his advisor. He showed that reduction is possible for any Diophantine number ρ, and number for which

$$\left|\rho - \frac{p}{q}\right| > \frac{c(\rho)}{q^{2+\delta}},$$

and obtained an optimal estimate of the smoothness of the transformation T_g as a function of δ.

Achieving these results required new ideas and an analytic technique that is important in the multi-dimensional case, which has not yet been solved.

2. Holomorphic dynamics.

I now turn to holomorphic dynamics. In it the study of the sequence of mappings of complex sets leads to a series of problems close to the theory of dynamical systems.

A typical problem of holomorphic dynamics is to describe the limiting sets of points of a mapping: $z \mapsto R(z)$, where $R(z)$ is a rational function, $z \in C$ or $\overline{C}$ (the Riemann sphere). Even the study of the sequence of iterations of such a seemingly simple mapping as $f_c(z) = z^2 + c$ conceals highly nontrivial results. To convey an idea regarding the structure of the set $\{f_c^n(z)\}$ I introduce two important concepts: the *Fatou set* and the *Julia set.* The Fatou set $F(f)$ of a mapping f is the set of regular points of the iterations f^n, i.e., the points z such that the family $\{f^n\}_{n=0}^{\infty}$ is equicontinuous in a neighborhood of the point z and the trajectories of the point z are Lyapunov stable. The Julia set $J(f)$ is the complement $\overline{C} \setminus F$. Thus the Riemann sphere decomposes into two invariant subsets: an open set F consisting of the regular points of the iterations f^n (the asymptotically stable points), and a closed set $J = \overline{C}^n \setminus F$ on which the behavior of the trajectories is stochastic. The study of Fatou and Julia sets for various classes of mappings $f(z)$ is a central problem of holomorphic dynamics.

Let us now return to the mapping

$$f_c(z) = z^2 + c. \tag{22}$$

For $c = -3$ the set $J(f_c)$ is a Cantor set, while for c small but nonzero the Julia set is a Jordan curve having no tangent at any point. The Julia set for the function $f_c(z) = z^2 - 1$ is a curve that divides the plane into a set of countable components. Computer models of the Julia sets yields pictures of marvelous beauty [PR].

In the late nineteenth and early twentieth centuries studies of dynamics on the Fatou set developed in parallel with the theory of ordinary differential equations, although specialists in each of the two fields often did not suspect the existence of those in the other. However, by the 1930s and 1940s the connection of these results was recognized, for example in the 1942 paper of Siegel, "On the reduction of an analytic transformation to a rotation in a neighborhood of a singular point." Siegel's result admits a natural interpretation in terms of differential equations (see the discussion on the first topic above). The corresponding efforts of Bryuno and Yoccoz also apply to it.

Although the studies of the Fatou set in the papers of A. Douady and J.H. Hubbard, Sullivan, and Thurston during the past decade have produced a rather complete description of it, knowledge of the structure of the Julia set remains far from complete for many classes of endomorphisms. Yoccoz solved a major problem connected with the structure of the Julia set for the mapping (22). The Julia sets J_c for mappings of the form (22) are either connected or have the structure of a Cantor set. Let J_c be connected. Consider the set M_c of values of the parameter c at which J_c is connected. This set is called the Mandelbrojt set. Douady and Hubbard [DH] proved that if J_c is connected, then M_c is also connected. Their paper showed that a more detailed description of the sets J_c and M_c depends on conditions for them to be locally connected. Yoccoz found a criterion for local connectedness of the sets J_c and M_c. Remarkably, the condition for local connectedness is closely linked with the question of whether it is possible to linearize mappings of the form $f_\alpha(z) = z^2 + e^{2\pi i \alpha} z$ in a neighborhood of zero. This question can be solved by the Bryuno-Yoccoz criterion.

I must now leave this beautiful area of modern mathematics, where the ideas and methods of so many of its different branches are concentrated in a small space: everything from the theory of Kleinian groups to the theory of approximation of real numbers, from the Teichmüller space to ordinary differential equations and computer graphics, an area in which history and modernity are strikingly intertwined [Mc1,2]. The unity of mathematics is shown best with these seemingly simple yet extraordinarily complicated examples.

The recently published title *Fields Medalist Lectures* presents surveys on the original articles of more than 20 Fields Medalists. Some of the papers relate to more current interests of the medalists [FM].

References

Remarks on the Literature

1. The proceedings of the congresses are the main source for detailed references. Nearly all laureates have delivered plenary or sectional talks, some more than once. Their talks are printed in the proceedings of the congresses and present a picture of the results obtained.

2. A beautifully edited illustrated history of the mathematical congresses has recently appeared [Al]. It briefly describes all the mathematical congresses starting with the 1893 congress in Chicago and contains photographs of all the Fields laureates and brief biographies of them. The book is carefully edited, though a few comic errors did creep through in biographical details of Soviet medalists.

3. An article of H. Tropp [Tr] offers basic material on the founding of the medal, including the text of Fields' will. The late French mathematician J. Dieudonné devoted several books [Di1–4] to an historical survey of the achievements in twentieth-century mathematics.

4. A conference on "Mathematical Research Today and Tomorrow" was held in Barcelona in 1991. Seven Fields medalists gave papers: A. Connes, G. Faltings, V. Jones, S.P. Novikov, S. Smale, R. Thom, and S.T. Yau. The proceedings of the conference are of great scientific interest, since they convey the views of leading mathematicians both on mathematics as a whole and on their own scholarly work [MR].

Literature

[Ah] L.V. Ahlfors. *Collected Papers*, 2 vol. Boston: Birkhäuser, 1982.

[Al] D.J. Albers, G.L. Alexanderson, and C. Reid. *History of Mathematical Congresses.* New York: Springer, 1986.

[AG] L. Alvarez-Gaumé. "Supersymmetry and the Atiyah-Singer index theorem." *Comm. Math. Phys.*, **90** (1983), 161–173.

[AGV] V.I. Arnol'd, S.M. Gusein-Zade, and A.N. Varchenko. *Singularities of Differentiable Maps*, 2 vol. Boston: Birkhäuser, 1985, 1988 (translated from Russian).

[At] M.F. Atiyah. *Collected Works*, 5 vol. Oxford: Clarendon Press, 1988.

[Ba] A. Baker, ed. *Advances in Transcendental Numbers.* Cambridge: Cambridge University Press, 1988.

[BPZ] A.A. Belavin, A.M. Polyakov, and A.B. Zamolodchikov. "Infinite conformal symmetries in two-dimensional quantum field theory." *Nucl. Phys.*, **B241** (1984), 333–380.

[Be] J. Bellisard. "Ordinary quantum Hall effect and noncommutative cohomology," in *Proc. Conf. on Localization of Disordered Systems*, W. Weller and P. Ziesche, eds., Leipzig: Teubner, 1988.

[Bi] J. Birman. "New points of view on knot theory." *Bull. Amer. Math. Soc.*, **28**, 2 (1993) 253–287.

[Bl] S. Bloch. "The proof of the Mordell conjecture." *Math. Intelligencer*, **6**, 2 (1984), 41–47.

[BT] E. Bombieri and J.E. Taylor. "Quasicrystals, tiling, and algebraic number theory: some preliminary convictions," in *The Legacy of Sonya Kovalevskaya*, L. Keen, ed., 241–264. Providence: American Mathematical Society, 1987.

[Con] A. Connes. *Noncommutative Geometry.* New York: Academic Press, 1994.

[CS] J.H. Conway and N.J.A. Sloane. *Sphere Packings, Lattices, and Groups*, 2nd ed. New York: Springer, 1993.

[CIL] M.G. Crendal, H. Ishii, and P.-L. Lions. "User's guide to viscosity solutions of second-order differential equations." *Bull. Amer. Math. Soc.*, **27**, 1 (1992) 1–67.

[Da] H. Davenport. *Multiplicative Number Theory*, 2nd ed. New York: Springer, 1980.

[DM] P. Deligne and G.D. Mostow. *Commensurabilities among Lattices in* $PU(1,n)$. Ann. Math. Studies, **132**. Princeton: Princeton University Press, 1993.

[Di1] J.A. Dieudonné. *A History of Algebraic and Differential Topology 1900–1960.* Boston: Birkhäuser, 1989.

[Di2] J.A. Dieudonné. *History of Functional Analysis.* Amsterdam: North-Holland, 1981.

[Di3] J.A. Dieudonné. *A Panorama of Pure Mathematics as seen by N. Bourbaki.* New York: Academic Press, 1982.

[Di4] J.A. Dieudonné. "The Beginnings of Italian Algebraic Geometry," in *MAA Studies in the History of Mathematics*, E. Phillips, ed., 278-299. Washington, D.C.: MAA, 1987.

[DH] A. Douady and J.H. Hubbard. "On the dynamics of polynomial-like mappings." *Ann. Sci. Ecole Norm. Sup. (4)*, **18** (1985), 287–344.

[Dr1] V.G. Drinfel'd. "Proof of the global Langlands conjecture for $GL(2)$ over a functional field." *Funkts. Anal. Pril.*, **11**, 3 (1977), 74–75 (in Russian). English transl.: *Funct. Anal. Appl.*, **11**, 3, 223–224.

[Dr2] V.G. Drinfel'd. "Quantum Groups," in *Proc. Berkeley IMC*, A. Gleason, ed. Providence: American Mathematical Society, 1987.

[DS] V.G. Drinfel'd and V.V. Sokolov. "Lie algebras and KdV-type equations." *Sovr. Probl. Mat. Nov. Dost.* 1984, (in Russian). English transl.: *J. Sov. Math.*, **30** (1985), 1975–2035.

[DFN] B.A. Dubrovin, A.T. Fomenko, and S.P. Novikov. *Modern Geometry*, 3 vol. Berlin: Springer, 1985–1988.

[Dy] F.J. Dyson. "Missed opportunities." *Bull. Amer. Math. Soc.*, **78** (1972), 635–652.

[Fa] L.D. Faddeev, N.Y. Reshetikhin, and L.A. Takhtadjan. "Quantization of Lie Groups and Lie Algebras." *Alg. Anal.*, **1** (1989), 178–206 (in Russian). English transl.: *Leningrad Math. J.*, **1** (1990), 193–225.

[Fal] G. Faltings. "The proof of Fermat's Last Theorem by R. Taylor and A. Wiles." *Not. AMS*, **42**, 7 (1995), 743–746.

[Fe] C. Fefferman. "The multiplier problem for the ball." *Ann. Math.*, **94**, 2 (1971) 330–336.

[FKV] S.M. Finashin, M. Kreck, and O.Y. Viro. "Exotic knottings of surfaces in the 4-sphere." *Bull. Amer. Math. Soc.*, **17**, 2 (1987), 287–290.

[Fl] A. Floer. "An instanton invariant for 3-manifolds." *Comm. Math. Phys.*, **118** (1988), 215–240.

[FM] M. Atiyah and D. Iagolnitzer, eds. *Fields Medalist Lectures*. Singapore: World Scientific, 1997.

[FU] D.S. Freed and K.K. Uhlenbeck. *Instantons and Four-Manifolds*, 2nd ed. New York: Springer, 1988.

[FH] M.H. Freedman and Z.-X. He. "Divergence-free fields: energy and asymptotic crossing numbers." *Ann. Math.*, **134** (1991), 189–229.

[FHW] M.H. Freedman, Z.-X. He, and Z. Wan. "Möbius energy of knots and unknots." *Ann. Math.*, **139** (1994), 1–50.

[FLM] I. Frenkel, J. Lepowsky, and A. Meurman. *Vertex Operators and the Monster*. New York: Academic Press, 1989.

[FKO] H. Furstenberg, Y. Katznelson, and D. Orenstein. "The ergodic-theoretic proof of Szemerédi's theorem." *Bull. Amer. Math. Soc.*, **7** (1982), 527–552.

[GKZ] I.M. Gelfand, M.M. Kapranov, and A.V. Zelevinsky. *Discriminants, Resultants, and Multidimensional Determinants*. Boston: Birkhäuser, 1994.

[GL] A.O. Gel'fond and Y.V. Linnik. *Elementary Methods in the Analytic Theory of Numbers*. Oxford: Pergamon Press, 1966.

[Gol] E. Golod. "On some problems of Burnside type," in *Proc. Int. Math. Cong.*, I. G. Petrovsky, ed. Moscow: Mir, 1968, 284–289.

[Gor] D. Gorenstein. *Finite Simple Groups.* New York: Plenum Press, 1982.

[GM] M. Goresky and R. Macpherson. *Stratified Morse Theory.* Berlin: Springer, 1989.

[Grom] M. Gromov. "Hyperbolic groups," in *Essays in Group Theory*, S. Gersten, ed., 75–263. Berlin: Springer, 1987.

[Grot1] *The Grothendieck Theory of Dessins d'Enfants*, L. Schneps, ed. Cambridge: Cambridge University Press, 1994.

[Grot2] *The Grothendieck Festschrift: a Collection of Articles*, 3 vol. P. Cartier, I. Illusie, N.M. Katz, G. Laumon, Y. Manin, K.A. Ribert. eds., 86–88. Boston: Birkhaüser, 1990.

[GKS] V. Guillemin, B. Kostant, and S. Sternberg. "Douglas' solution of the Plateau problem." *Proc. Nat. Acad. Sci. USA*, **185** (1988), 3277–3278.

[HR] H. Halberstam and K.F. Roth. *Sequences*, Vol. 1. Oxford: Clarendon Press, 1966.

[He1] D. Hejhal. "The Selberg trace formula for PSL $(2, \mathbb{R})$." I. *Lect. Notes Math.*, No. 548. Berlin: Springer, 1976.

[He2] D. Hejhal. "The Selberg trace formula for PSL $(2, \mathbb{R})$." II. *Lect. Notes Math.*, No. 1001. Berlin: Springer, 1983.

[Hi] N.J. Hitchin. "Monopoles and Geodesics." *Commun. Math. Phys.*, **83** (1982), 579–602.

[HG] L. Hörmander and L. Gårding. "Why is there no Nobel Prize in mathematics?" *Math. Intelligencer*, **7**, 3 (1985), 73–74.

[Hö] L. Hörmander. *The Analysis of Linear Partial Differential Operators*, 4 vol. New York: Springer, 1985.

[Je] R. Jensen. "The maximum principles of viscosity solutions of fully nonlinear elliptic particle differential equations." *Arch. Rat. Mech. Anal.*, **101** (1988), 1–27.

[J1] V.F.R. Jones. "A new knot polynomial and von Neumann algebras." *Not. Amer. Math. Soc.*, **33** (1986), 219–225.

[J2] V.F.R. Jones. "Subfactors and knots." *CBMS*, No. 80. Providence: American Mathematical Society, 1991.

[J3] V.F.R. Jones. "Knot theory and statistical mechanics." *Scientific American*, **263**, 5 (1990), 98–103.

[Jo] D. Joravsky. *The Lysenko Affair.* Chicago: University of Chicago Press, 1986.

[Kac] V. Kac. *Infinite-dimensional Lie Algebras*, 3rd ed. Cambridge: Cambridge University Press, 1994.

[Kat] N.M. Katz. "An overview of Deligne's proof of the Riemann hypothesis for varieties over finite fields," in *Mathematical Developments Arising from the Hilbert Problems, Proc. Symp. Pure Math.*, F. Browder, ed., Vol. 28, part 1, 275-305. Providence: Amer. Math. Soc., 1976.

[Ke] M. Kervaire. "A manifold which does not admit any differentiable structure." *Comm. Math. Helv.*, **34** (1969), 257–270.

[Kod] K. Kodaira. *Collected Works*, 3 vol. Princeton: Princeton University Press, 1975.

[Kol1] J. Kollár. "The structure of algebraic threefolds: an introduction to Mori's program." *Bull. Amer. Math. Soc.*, **17**, 2 (1987), 211–273.

[Kol2] J. Kollár, ed. "Flips and abundance for algebraic threefolds." *Astérisque*, 211 (1992).

[Kos] A.I. Kostrikin, "Sandwiches in Lie algebras." *Mat. Sb.*, **110** (1979), 3–12 (in Russian). English transl.: *Math. USSR-Sb.*, **38** (1981), 1–9.

[L1] H.B. Lawson, Jr. *Lectures on Minimal Submanifolds.* Berkeley: Publish or Perish, Inc., 1980.

[L2] H.B. Lawson, Jr. "The Theory of Gauge Fields in Four Dimensions." *CBMS*, No. 58. Providence: American Mathematical Society, 1985.

[Lin] J. Lindenstrauss. "Jean Bourgain." *Not. Amer. Math. Soc.*, **41**, 9 (1994), 1103–1105.

[Lio1] P.-L. Lions. "On kinetic equations." *Proc. Int. Math. Cong.*, The Mathematical Society of Japan, 1173–1185. Tokyo: Springer, 1991.

[Lio2] P.-L. Lions, "On some recent method for nonlinear partial differential equations," in *Proceedings of the International Mathematical Congress*, S. Chaterji, ed., Vol. 1, 140–156. Basel: Birkhäuser, 1995.

[MR] *Mathematical Research Today and Tomorrow: Viewpoints of Seven Fields Medalists*, Lect. Notes. Math., No. 1525. New York: Springer, 1992.

[Marg1] G.A. Margulis. *Discrete Subgroups of Semisimple Lie Groups*. Berlin: Springer, 1991.

[Marg2] G.A. Margulis. "Oppenheim conjecture," in *Fields Medals Lectures*, M. Atiyah, D. Iagolnitzer, eds., 272–327. Singapore: World Scientific, 1997.

[Mart] D.A. Martin. "Hilbert's first problem: the continuum hypothesis," in *Mathematical Developments Arising from the Hilbert Problems, Proc. Symp. Pure Math.*, F. Browder, ed., Vol. 28, Part 1, 275–305. Providence: AMS, 1976.

[Mc1] C.T. McMullen. *Complex Dynamics and Renormalization.* Princeton: Princeton University Press, 1994.

[Mc2] C.T. McMullen. *Renormalization and 3-manifolds Rich Fiber Over the Circle.* Princeton: Princeton University Press, 1996.

[Mig] A. Migdal. "Quantum gravity as dynamical triangulation," in *Two-dimensional Quantum Gravity and Random Surfaces*, D.J. Gross, T. Piram, and S. Weinberg, eds., 41–79. Singapore: World Scientific Publishing, 1992.

[Mi1] J.W. Milnor. *Morse Theory.* Princeton: Princeton University Press, 1963.

[Mi2] J.W. Milnor. *Lectures on the h-cobordism Theorem.* Princeton: Princeton University Press, 1965.

[Mo] M.I. Monastyrsky, "The Dirac monopole and the Hopf invariant," in *Proceedings of the Legacy of P. Dirac*, B. Medvedev, ed., 66–76. Moscow: Nauka, 1990. (In Russian.)

[Po] A. van der Poorten. *Notes on Fermat's Last Theorem.* New York: Wiley & Sons, 1996.

[PR] H.-O. Peitgen and P.H. Richter. *The Beauty of Fractals.* Berlin: Springer, 1989.

[Q] D.G. Quillen, "Determinants of Cauchy-Riemann operators on Riemann surfaces." *Funct. Anal. Appl.*, **19**, 1 (1985), 31–34.

[Re] T. Regge. "On strings and J. Douglas' variational principle," in *Differential-geometric Methods in Theoretical Physics*, K. Bleuler and M. Werner, eds., 145-148. Dordrecht: Kluwer, 1988.

[Ri] P. Ribenboim. *Catalan's Conjecture: Are 8 and 9 the Only Consecutive Powers?* Boston: Academic Press, 1994.

[Ro] K.O. Rossianov. "Joseph Stalin and the 'New' Soviet Biology." *Isis*, **84**, 4 (1993), 728–745.

[RS] K. Rubin and A. Silverberg. "A report on Wiles' Cambridge lectures." *Bull. Amer. Math. Soc.*, **33**, 1 (1994) 15–38.

[SBGC] D. Schechtman, I. Blech, D. Gratias, and J.W. Cahn. "Metallic phase with long-range orientational order and no translational symmetry." *Phys. Rev. Lett.*, **53** (1984), 1951–1956.

[S1] J.-P. Serre. *Oeuvres*, 3 vol. Berlin: Springer, 1986.

[S2] J.-P. Serre. *Lie Algebras and Lie Groups.* New York: Benjamin, 1965.

[S3] J.-P. Serre. *Cours d'arithmétique.* Paris: Presses Universitaires de France, 1970.

[Sh] T.N. Shorey and R. Tijdeman. *Exponential Diophantine Equations.* Cambridge: Cambridge University Press, 1986.

[Sm1] S. Smale. "Differentiable dynamical systems." *Bull. Amer. Math. Soc.*, **73** (1967), 747–817. (Reprinted in [Sm2].)

[Sm2] S. Smale. *The Mathematics of Time.* New York: Springer, 1980.

[ST] D.P. Sullivan and N. Teleman. "An analytic proof of Novikov's theorem on rational Pontryagin classes." *IHES Publ. Math.*, 58 (1983), 79–81.

[Sza] L. Szapiro, ed. "Séminaire sur les pinceaux arithmétiques: la conjecture de Mordell." *Astérisque*, 127 (1985).

[Sze] E. Szemerédi. "On sets of integers containing no elements in arithmetic progression." *Acta Arith.*, **27** (1975), 199–145.

[Th] R. Thom. *Structural Stability and Morphogenesis.* Reading, MA: W.A. Benjamin, 1975.

[T1] W.P. Thurston. *Three-dimensional Geometry and Topology,* vol. I., S. Levi, ed. Princeton: Princeton University Press, 1997.

[T2] W.P. Thurston. "On the geometry and dynamics of diffeomorphisms of surfaces." *Bull. Amer. Math. Soc. (NS)*, **19**, 2 (1988), 417–431.

[TW] W.P. Thurston and J.R. Weeks. "The mathematics of three-dimensional manifolds." *Scientific American*, **251**, 1, (1984), 108–120.

[Tr] H.S. Tropp. "The origins and history of the Fields medal." *Historia Math.*, **3** (1976), 167–181.

[Va] V.A. Vasil'ev. "Invariants of knots and complements of discriminant." in *Developments in Mathematics. The Moscow School*, V. Arnol'd and M. Monastyrsky, eds., 194–250. London: Chapman and Hall, 1993.

[Ve] A.B. Venkov. "The spectral theory of automorphic functions." *Trudy Mat. Inst. Steklov*, **153** (1981) (in Russian). English transl.: *Proc. Steklov Inst. Math*, 4 (1982).

[We] H. Weyl. *Algebraic Theory of Numbers.* Princeton: Princeton University Press, 1960.

[Wil] A. Wiles. "Modular elliptic curves and Fermat's Last Theorem." *Annals of Math.*, **141** (1995), 443–551.

[Wi1] E. Witten. "Constraints on supersymmetry breaking." *Nucl. Phys.*, **B202** (1982), 253–316.

[Wi2] E. Witten. "Quantum field theory and the Jones polynomial." *Comm. Math. Phys.*, **121** (1989), 351–399.

[Wi3] E. Witten. "Algebraic geometry associated with matrix models of two-dimensional gravity," in *Topological Methods in Modern Mathematics,* L.R. Goldberg, A.V. Phillips, eds., 235–269. Houston: Publish or Perish, 1993.

[Za] L. Zalcman. "Mathematicians Sweep 1988 Wolf Prizes." *Math. Intelligencer*, **11**, 2 (1989), 39–48.

[Ze1] E.I. Zelmanov. "Solution of the restricted Burnside problem for groups of odd exponent." *Izv. Akad. Nauk SSSR*, Ser. Mat., **54** (1990), 42–59 (in Russian). English transl.: *Math. USSR-Izv.*, **36** (1991), 41–60.

[Ze2] E.I. Zelmanov. "Solution of the restricted Burnside problem for 2-groups." *Math. Sb.*, **183** (1991), 568–592. (in Russian). English transl.: *Math. USSR-Sb.*, **72** (1992), 543–565.

[Ze3] E.I. Zelmanov. "On the theory of Jordan algebras." in *Proc. Int. Math. Cong.*, 455-463. Amsterdam: North-Holland, 1984.

[Ze4] E.I. Zelmanov. "On periodic compact groups." *Isr. J. Math.*, **77** (1992), 83–95.

Index of Names